Science for Motor Vehicle Engineers

Science for Motor Vehicle Engineers

Peter Twigg

OXFORD AMSTERDAM BOSTON LONDON NEW YORK PARIS
SAN DIEGO SAN FRANCISCO SINGAPORE SYDNEY TOKYO

Butterworth-Heinemann
An imprint of Elsevier Science
Linacre House, Jordan Hill, Oxford OX2 8DP
225 Wildwood Avenue, Woburn MA 01801-2041

First published 1996
Transferred to digital printing 2002

British Library Cataloguing in Publication Data
A catalogue record for this book is available from the British Library

Library of Congress Cataloguing in Publication Data
A catalogue record for this book is available from the Library of Congress

ISBN 0 340 64527 X

For information on all Butterworth-Heinemann publications
visit our website at www.bh.com

Contents

Preface

This book has been written to give a background to the essential theories and applications of motor vehicle science. It is a suitable companion for readers following a wide range of motor vehicle courses, or as a text for automobile engineering students not following a formal course. All the engineering concepts have a specific motor vehicle application wherever possible. The aim is not to present abstract theories but to use realistic examples, useful to readers in their studies.

Engineering requires a certain knowledge of mathematics, but the mathematics here has been kept as simple as possible without sacrificing understanding. It is expected that as the reader works through the examples and problems, s/he will develop a greater understanding of the mathematics required for engineering. Great care has been taken in the explanation and calculation of units, as units tend to cause a great deal of confusion and mistakes in engineering, particularly mechanical engineering with its wide range of quantities and multiple factors. The concept of unity brackets has been followed throughout the text, as this is the only method that ensures correct analysis of the units through the calculations. Chapter 1 is an introduction to the book, covering the SI system, background mathematics and advice on problem solving, particularly exam questions. Chapters 2–5 deal with mechanics, both statics and dynamics. Chapter 6 covers several aspects of thermo-dynamics. Chapter 7 is an introduction to control and instrumentation. Chapters 8 and 9 cover metals and electricity.

To help the reader develop a sound grasp of the principles covered there are many diagrams, examples and problems as an aid to develop knowledge and understanding.

Problem answers are shown at the end of the book.

I should like to thank Christine, Unzie and mum for their help and constructive comments. Most of all thanks to Denise for all her support and patience.

1 An introduction to motor vehicle science

1.1 The SI System

Engineering quantities and units

In engineering, we have to make sense of various different quantities, such as force, mass, acceleration and temperature. The quantities need to be measured with standard units: for example, if a piston with the same dimensions made in Germany has a diameter of 100 millimetres then a similar piston made elsewhere should have the same actual width, otherwise engines would not fit together. In other words German millimetres must be the same as Italian and Japanese millimetres. An international agreement defines these various units. This agreement is called the Système International d'Unités, the international symbol for which is SI. We know this as the metric system.

Besides specifying a quantity by both its number and its unit symbol, in equations the quantity is replaced by an algebraic quantity symbol which can usually have any value. The primary quantities met with in this book are shown in Table 1.1.

All other quantities that we need can be derived from these primary quantities. For example, area can be measured as the product of two lengths. We can then say that with the unit length of the metre, the unit area is metres × metres which are called square metres (m^2). When solving engineering problems, you can enclose the unit symbol in square brackets to help to keep calculations clear and uncluttered.

In the section on mechanics all units used in measurement will be derived from length, mass and time. The quantities of length and time you are probably already familiar with. Mass will be discussed at an early stage in the book. The unit of temperature, the kelvin, is the absolute version of Celsius and will be used in the section on thermodynamics. Electrical current will not be required until the section on electricity. There are two other base units in the SI system (the candela and the mole), but these will not be of any interest to us in motor vehicle science.

Table 1.1

Quantity	Quantity symbol	Unit	Unit symbol
length	l	metre	[m]
mass	m	kilogram	[kg]
time	t	second	[s]
electric current	I	ampere	[A]
temperature	T	kelvin	[K]

Multiples and submultiples

Sometimes the standard units will be too large or too small for the required measurement. For example, you would not want to measure a spark plug gap in metres as the number would require more digits than is convenient (a typical spark plug gap is around 0.0006 m). To help simplify measurements and calculations the SI system uses prefixes with the base units. These are the same whichever base unit they are applied to. The ranges of multiples and submultiples you are likely to need in motor vehicle science are shown in Table 1.2.

By using the prefix milli in front of metres, the spark plug gap above may now be described by: 0.6 mm (= 0.6 × 0.001 m). You may come across some smaller submultiples in electronic components (nano = × 10^{-9} and pico = × 10^{-12}) but we won't look at those here. The four multiple and submultiples, hecto, deca, deci and centi should be avoided if possible as they do not multiply the base unit by 10 to the power of a multiple of three. This is important for simplifying calculations as you will see later.

Only one multiplying prefix is used at any one time to a given unit: a thousandth of a millimetre is not described as 1 milli-millimetre but as 1 micrometre (1 μm = 0.000 001 m). If you hear the word micron, this is a former name for a micrometre. Similarly, one thousand kilograms, is a megagram [Mg] and not a kilo-kilogram. You may have noticed that mass does not quite fit into the system in the same way as other quantities, as the base unit is the kilogram and not the gram as you would expect. This does not alter the way in which prefixes are used: masses are still expressed as multiples of a gram. So 10^{-6} kg should still be written as 1 mg. When a prefix is applied to a unit it becomes part of that unit and is subject to any mathematical functions applied. For instance, to express 1 mm³ in terms of metres:

$$1 \text{ mm} = 10^{-3} \text{ m}$$

Remember when working with equations that what you do to one side of the equals sign you must do to the other. Here, we can raise each side to the power of three:

$$(1 \text{ mm})^3 = (10^{-3} \text{ m})^3$$

Powers inside and outside the brackets multiply to give:

$$1^3 \text{ mm}^3 = 10^{-9} \text{ m}^3$$
$$1 \text{ mm}^3 = 10^{-9} \text{ m}^3$$

Table 1.2

Prefix	Symbol	Factor by which unit is multiplied	
giga	G	10^9	1 000 000 000
mega	M	10^6	1 000 000
kilo	k	10^3	1 000
hecto	h	10^2	100
deca	da	10^1	10
deci	d	10^{-1}	0.1
centi	c	10^{-2}	0.01
milli	m	10^{-3}	0.001
micro	μ	10^{-6}	0.000 001

Now look at this example.

An engine exhaust valve has a mass, m, of 182 g and a volume, V, of 22957 mm^3. Find the density of the material.

$$182 \text{ g} = 182 \times 10^{-3} \text{ kg}$$

and

$$22957 \text{ mm}^3 = 22957 \times 10^{-9} \text{ m}^3.$$

The density, which has the quantity symbol ρ, can be calculated as follows:

$$\rho = \frac{m}{V} = \frac{182 \times 10^{-3}}{22957 \times 10^{-9}} \left[\frac{\text{kg}}{\text{m}^3}\right]$$

When a number to the power of a minus number appears as a multiple in the denominator of a calculation, it may be changed to a multiple of the numerator by raising it to the power of the same number in a positive form. In this example the 10^{-9} in the denominator can be moved to the numerator and changed to 10^9 (see Chapter 1.3 on background maths).

$$\rho = \frac{m}{V} = \frac{182 \times 10^{-3} \times 10^9}{22957} \left[\frac{\text{kg}}{\text{m}^3}\right]$$

The power multiples of three can then be added together like this,

$$10^{-3} \times 10^9 = 10^6$$

to give:

$$\rho = \frac{m}{V} = \frac{182 \times 10^6}{22957} \left[\frac{\text{kg}}{\text{m}^3}\right] \quad = 7927.86 \text{ kg/m}^3$$

By using 10 to the power of multiples and submultiples of three with base units throughout problems, calculations are made simpler and the answer is in a form ready for a unit prefix. Notice that the units of the answer appear automatically. It is easy to make a mistake in engineering calculations with all the different units used. By incorporating the units into the initial calculation and following them through, the correct units should appear with the answer. Units are multiplied and divided in the same way as numbers. The standard units of density are kg/m^3. Had the expression been copied down wrongly, for instance as $\rho = V/m$, then the units of the answer would have been m^3/kg and the mistake would have been spotted immediately.

Here is another example:

The sides of a rectangular fuel tank measure 600 mm by 30 cm by 0.5 m. The volume of the tank can be found by multiplying the three sides together. Let's start by converting the units into standard form rather than trying to do it within the calculation. The standard unit for length is the metre.

$$600 \text{ mm} = 600 \times 10^{-3} \text{ m}$$
$$30 \text{ cm} \quad = 30 \times 10^{-2} \text{ m}$$

0.5 m is already in standard form.

$$V = 600 \times 10^{-3} \text{ m} \times 30 \times 10^{-2} \text{ m} \times 0.5 \text{ m}$$

(add the -3 and -2 together, i.e. $10^{-3} \times 10^{-2} = 10^{-5}$)

$$= 600 \text{ m} \times 30 \text{ m} \times 0.5 \text{ m} \times 10^{-5}$$
$$= 9000 \times 10^{-5} \text{ m}^3 = 90 \times 10^{-3} \text{ m}^3$$

Remember that this does not equal 90 mm^3, because 1 mm^3 does not equal 10^{-3} m^3, (1 mm^3 = 10^{-9} m^3). There are 1000 litres in 1 m^3 though, so we can say that this equals 90 litres. Again the units provide a good check. If the unit of the answer was, for instance, m^2 a mistake would have been made, since this is the unit of area and not volume.

The imperial system and unit conversion

For many years in the UK the imperial system was used. Many vehicles around are dimensioned in imperial units. There are also many people who still use imperial units. For this reason you need to be able to convert units from one system to the other.

The imperial system is based on the foot, pound and second. The foot [ft] is one-third of the imperial yard which is defined as 0.9144 metre. The pound [lb] is now defined as 0.4536 kilograms.

For example, if a mass is given as 25 lb, then to convert this to kilograms you would multiply this by 0.4536, as follows:

$$25 \, [\cancel{\text{lb}}] \times 0.4536 \, [\text{kg}/\cancel{\text{lb}}] = 11.34 \, \text{kg}$$

Notice how the [lb] units cancel leaving [kg] for the answer. To carry out the reverse of this the reciprocal of 0.4536 kg/lb is used:

$$\frac{1}{0.4536 \, [\text{kg/lb}]} = 2.2046 \, \text{lb/kg}$$

The units of the answer are now also inverted. This method can be used for converting kilograms to pounds:

$$11.34 \, [\cancel{\text{kg}}] \times 2.2046 \, [\text{lb}/\cancel{\text{kg}}] = 25 \, \text{lb}$$

Again notice how the [kg] units cancel leaving [lb] with the answer. The tables of values for converting from one system to another are commonplace: e.g. wall-charts, front diaries etc. but mistakes are easily made. It is sometimes not obvious whether one conversion value should be used, or its reciprocal. Here, units play a major role. Had we tried to convert 11.34 kg to pounds using 0.4536 kg/lb, then the answer would have been incorrect:

$$11.34 \, [\text{kg}] \times 0.4536 \, [\text{kg/lb}] = 5.144 \, [\text{kg}] \times [\text{kg/lb}] \qquad (wrong)$$

The units produced with the answer would not have been [lb] as the [kg] units are both numerators and would not cancel. Table 1.3 shows some common conversion factors between SI and imperial units.

The conversion values 0.4536 kg/lb and 2.2046 lb/kg are known as unity brackets. These are a very useful concept, not only when converting between the SI system and imperial system but when converting from any one unit to another, e.g. m^3 to mm^3.

The actual value of a unity bracket may seem strange when there is a number present other than one. This means that any quantity can be multiplied by a unity bracket to change the units without altering the value. Using unity brackets will seem rather awkward at first but, once you are familiar with them, they will save time and prevent any confusion when converting units. They are easy to calculate and it is

Table 1.3

Imperial unit	SI unit
1 inch	25.4×10^{-3} m
1 foot	0.3048 m
1 yard	0.9144 m
1 gallon	$4.546\,09 \times 10^{-3}$ m^3 (note 1 gallon = 8 pints)
1 pint	$0.5682\,613 \times 10^{-3}$ m^3
1 pound	0.453 59 kg
1 ton	$1.016\,047 \times 10^{-3}$ kg
1 mph	0.447 04 m/s
1 horse power	745.7 W
1 mile	$1.609\,344 \times 10^3$ m

better for you to know how to work them out for yourself rather than having a ready-made list of them. Consider the units of feet and metres.

$$1 \text{ ft} = 0.3048 \text{ m}$$

Here it is important to realise that the unit symbol [ft] is not fixed to the number 1. Both sides are now divided by [ft].

$$1\left[\frac{\text{ft}}{\text{ft}}\right] = 0.3048\left[\frac{\text{m}}{\text{ft}}\right]$$

The two [ft] symbols on the left cancel out leaving the value of one, i.e. a unity bracket:

$$1 = 0.3048\left[\frac{\text{m}}{\text{ft}}\right]$$

This can be inverted if necessary and will still equal one and therefore still be a unity bracket:

$$1 = 0.3048\left[\frac{\text{m}}{\text{ft}}\right] = \frac{1}{0.3048}\left[\frac{\text{ft}}{\text{m}}\right]$$

You now have two forms of a unity bracket that can be used for converting feet to metres or metres to feet as follows:

$$15\,[\text{ft}] \times 0.3048\left[\frac{\text{m}}{\text{ft}}\right] = 4.572\,[\text{m}]$$

$$4.572\,[\text{m}] \times \frac{1}{0.3048}\left[\frac{\text{ft}}{\text{m}}\right] = 15\,[\text{ft}]$$

Again had we tried to use the wrong version of the unity bracket the units of the answer would not be correct. Here's another one:

$$1\,[\text{mm}^3] = 10^{-9}\,[\text{m}^3]$$

$$1 = 10^{-9}\left[\frac{\text{m}^3}{\text{mm}^3}\right] = \frac{1}{10^{-9}}\left[\frac{\text{mm}^3}{\text{m}^3}\right]$$

We can now use this to convert [mm³] to [m³] or [m³] to [mm³]. Convert 5000 mm³ to m³.

$$5000 \, [\text{mm}^3] \times 10^{-9} \left[\frac{\text{m}^3}{\text{mm}^3} \right] = 5 \times 10^{-6} \, [\text{m}^3]$$

Notice how the unit we want to get rid of is always on the bottom of the unity bracket and the unit we do want is on the top.

$$5 \times 10^{-6} \, [\text{m}^3] \times \frac{1}{10^{-9}} \left[\frac{\text{mm}^3}{\text{m}^3} \right] = 5000 \, [\text{mm}^3]$$

Don't worry if unity brackets are a little unclear at this stage. After a bit of practice these will present no problem to you at all. Try working through the following problems before continuing.

Problems 1.1

1. If a petrol tank holds 20 gallons, how much does it hold in cubic metres and litres?
2. If the engine capacity of a bike is 1200 cc, what is this in m³ (cc stands for cubic centimetres)?
3. An engine block has a mass of 250 lb, what is this in kg?
4. If the width of a car is 63 inches, what is this in m?
5. If the mass of a car is 1784 lb, what is this in kg?
6. If the turning circle of a car is 32.75 ft, what is this in m?
7. If an engine requires 7 pints of lubricating oil, how much is this in litres and in m³?
8. If the wheel base of a car measures 91.3 inches, what is this in m?
9. The valve clearance on the inlet to a car engine measures 0.006 inches with imperial feeler gauges. The workshop manual specifies that 0.12 to 0.17 mm are required. Is the current clearance acceptable?
10. A car travels at 40 mile/h. What is its speed in metres per second?

1.2 Tackling problems

Exam questions

It is assumed that at some stage you will have to solve both theoretical exam questions and real-life problems. Exam questions are by far the most difficult. You are usually being pushed for time, and tension and nerves do not help you to think very clearly. You will have far more success if you work through the questions carefully and methodically rather than rushing through them as quickly as possible. Admittedly, this is easier said than done but practice will enable you to break a problem down into manageable steps. Often, you will be given a choice of questions and it is best to read through them all carefully before deciding which to tackle first. As a guide to time, divide the total exam time by the number of questions you have to answer. When you have spent your allocated time on one question, leave it and go on to the next one. It is far better to have two questions two-thirds finished than one fully completed, and most of the easier marks for a question tend to be at the beginning of it. You can always go back and finish off questions if you have time later.

On first reading through a problem, you will be presented with a lot of information at once and it is difficult to visualise what is happening. Always start by drawing a simple sketch and writing quantities where possible on the drawing.

Look at this example.

The bore of a car engine is 72 mm and the stroke is 82 mm. The clearance volume of each cylinder when the piston is at top dead centre is 45 117 mm^3. Find the compression ratio.

On reading this, it may seem like a lot of numbers are thrown at you with little meaning. So, start with a simple freehand sketch like the one shown here.

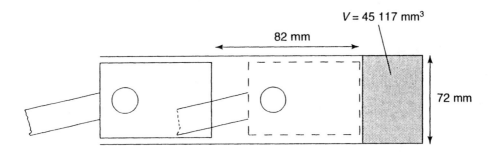

Don't worry about detail, scale and having to use a ruler. Now the information is in a more understandable form. Leave the rest of this question as we'll look at compression ratios later on in the book. The method required to answer a question is not always apparent and in an exam it is easy to panic. A simple diagram clarifies the situation and you will have a better idea about how to tackle the problem. Always show all of your calculations no matter how trivial. A question with the wrong answer can still pick up most of the marks for the method of solving it, provided that the calculations are shown. Marks are awarded for each stage and not just the final answer.

Electronic calculators

If you do not currently use a calculator then you should buy one. They are cheap, compact, powerful and reliable. You probably know how to use the main functions such as add, multiply and divide. It is well worth learning how to use some of the lesser known functions for engineering calculations as well. This will save you a lot of time and mistakes. In this section, a calculator function key is represented by brackets, e.g. the add button would be (+).

Exponent input (EXP): This is the most useful key to an engineer. It is usually written as EXP on the button and is short for exponent. This is for multiplying a number by 10 to the power of another number. It is used like this: 555 Mg = 555 × 10^3 kg and this number can be entered into the calculator by pressing 555 (EXP) 3. The way of making a number negative varies but usually there is a button with a − sign on it that is pressed before or after the relevant number. So 235 mg = 235 × 10^{-3} kg and can be entered by pressing 235 (EXP) (−) 3. Sometimes in data books this would be written as 235E − 3 and calculators often give answers in this form.

Power key (x^y): This is used for raising one number to the power of another number. To calculate $19^{1.3}$, press 19 (x^y) 1.3. This can be used in the same way as the (EXP) key but not as quickly: 555 × 10^3 can be entered as 555 (×) 10 (x^y) 3. This process can be reversed usually by a separate button ($_x\sqrt{}$), or by using an inverse button marked (INV) or (2nd) before the (x^y), key depending on the type of calculator. So 19 (x^y) 1.3 (=) 45.96. This is reversed as follows: $_{1.3}\sqrt{45.96}$ = 19 can be calculated by pressing 45.96 (INV) (x^y) 1.3 (=) 19.

Engineering key (ENG): The answer to some calculations will be given by the calculator in an exponential form. By pressing this key you can convert the exponent to a multiple of three so that it can easily be expressed by the SI system prefixes. For example, if the calculator showed 45.8E − 5 (= 45.8 × 10^{-5}), then by pressing (ENG) this would be changed to 458E − 6, which has the same value but can be expressed using the micro [μ] prefix. By pressing (INV) (ENG) the reading would be changed to 0.458E − 3 which also has the same value but can easily be expressed using the prefix milli [m].

Reciprocal key (1/x) or (x^{-1}): This is another useful function. It simply gives the reciprocal of a value. For example, 1/47.5 could be calculated by pressing 47.5 (1/x).

The precise way in which these functions are used varies from calculator to calculator but this should give you the general idea. Other useful functions are (sin), (cos), (tan) and (DRG) but these will be referred to as they are needed in later sections.

1.3 Mathematics

In the presentation of theories in this book the aim has been to keep the level of mathematics as simple as possible, without losing any clarity of the explanations. Mathematics is one of the principal tools of an engineer and a certain level of mathematics is necessary for all studies of engineering. Algebraic manipulation of some sort is particularly necessary for most subjects. This chapter is intended to be a brief revision of mathematical techniques necessary for understanding the theories in this book. If you feel confident in basic algebra and such things as basic trigonometry, then skip this section.

Algebra

There is sometimes an air of mystery surrounding algebra, and often this is created by the best mathematicians. Someone who can carry out algebraic calculations easily, without thinking, will perhaps have difficulty in explaining it to someone else. Algebra is very simple though, if you stick to a simple set of rules.

Consider a mathematical expression written with specific numbers,

e.g. $3 + 4 = 4 + 3$

We are saying here that it does not matter whether four is added to three or whether three is added to four: the answer is the same. This is not only true of the numbers three and four though, it is true for any number. One way of showing this is with algebra. Pick any two symbols you like to represent the two numbers. How about the letters x and y? The above expression could be rewritten as follows:

$x + y = y + x$

The expression can now be applied to any numbers.

This is the principle behind algebra: symbols are used to represent any magnitude of a quantity.

If a vehicle can travel 30 miles on one gallon of fuel, then we could write this in algebraic terms as:

$M = 30 F$

where F represents the fuel quantity in gallons and M represents the distance travelled in miles. This is usually referred to as a formula. If we have four gallons of fuel, then $F = 4$. The miles travelled, M, can then be calculated using the formula:

$M = 30 \times 4 = 120$ miles

Some formulae (plural of formula) in engineering may look very complex, but each letter or symbol is used to represent any number of a particular quantity, such as F and M above. Notice also that the multiplication symbol $\times$ is missed out. If two components of an expression are directly next to each other then they should be multiplied. Sometimes a dot is used in place of the multiplication symbol. Quantities such as M or F that can vary are sometimes referred to as **variables**.

The usual operations that can be applied to numbers can also be applied to any algebra symbols. For example:

$y = 4b^3$

$\quad = 4 \times b \times b \times b$

If $b = 5$ then $y = 4 \times 5 \times 5 \times 5 = 500$

If parts of an equation appear in brackets then work out these parts of the calculation first. For example

$s = \left(\dfrac{a}{b} + c\right)^2$

The a must be divided by the b and then the c added before any squaring is done.

Therefore $s = \left(\dfrac{a}{b} + c\right) \times \left(\dfrac{a}{b} + c\right)$

not $\dfrac{a^2}{b^2} + c^2$ as this gives a completely different answer.

If a square root sign is present, then all the algebra symbols underneath the line must be calculated first.

$$t = \sqrt{d\,e\,f}$$

Multiply d and e and f first, and then find the square root of that answer for the final answer. Besides the square root, other roots are sometimes required.

$$\text{If } r = \sqrt[4]{V}$$

$$\text{then } V = r \times r \times r \times r$$

Roots are sometimes represented in a different way.

$$\text{If } r = \sqrt[4]{V}$$

$$\text{then also: } r = V^{(\frac{1}{4})} = V^{(0.25)}$$

So if $V = 7$, then

$$r = \sqrt[4]{7}$$

Try both ways on a calculator:

$$r = 7\,(\sqrt[x]{y})\,4 \quad = 1.6256$$

$$r = 7\,(y^x)(1/4) = 1.6256$$

If a number is raised to the power of a negative number, then treat it as a positive number and invert the answer.

$$\text{If } w = Q^{-2.3}$$

$$\text{this is the same as saying } w = \frac{1}{Q^{2.3}}$$

$$\text{Or if } p = R^{-0.5}$$

$$\text{this is the same as saying } p = \frac{1}{R^{0.5}}$$

$$\text{Also as } 0.5 = \frac{1}{2} \text{ then } p = \frac{1}{\sqrt[2]{R}}$$

❑ Example 1.1

If $a = 4$ and $b = 7.9$, find c for the expression below.

$$c = 15(b^{-a} + 0.009)$$

Put the numbers into the equation:

$$c = 15(7.9^{-4} + 0.009)$$

$$= 15\left(\frac{1}{7.9^4} + 0.009\right)$$

$$= 15\left(\frac{1}{3895.008} + 0.009\right)$$

$$= 15 \times 0.009\,257 = \mathbf{0.1388}$$

❏ Example 1.2

Find a, if $N = 24$ and $k = 2$.

$$a = 13 + (N \times 10^{-k})$$

Put the numbers into the equation:

$$a = 13 + (24 \times 10^{-2})$$

$$= 13 + \left(24 + \frac{1}{10^2}\right)$$

$$= 13 + (24 \times 0.01)$$

$$= 13 + 0.24 = \mathbf{13.24}$$

One variable can be said to be proportional to another variable rather than equal to it. The symbol for proportional is $\propto$ and this is used in place of the equal sign:

e.g. $y \propto x$

y is proportional to x. This means that y is equal to x multiplied by a constant value but we do not know what this constant value is. If the constant was represented by C, then

$$y = x \times C$$

Transposition of formulae

Transposition of a formula means to rearrange it to a more suitable form. For example, if

$$U = 3.9B + 10$$

and we know the value of U and require the value of B, then the formula is not suitable in its present form. We can carry out any operation on a formula, provided that we do exactly the same thing to each side of the equal sign: e.g. add seven to each side, multiply each side by 100, square both sides, anything you like. The equation will still balance. This is a useful technique for rearranging a formula. In the above equation, we could start by subtracting 10 from each side:

$$U - 10 = 3.9B + \cancel{10} - \cancel{10}$$

As the tens on the right cancel, it leaves:

$$U - 10 = 3.9B$$

If we now divide each side by 3.9:

$$\frac{U - 10}{3.9} = \frac{\cancel{3.9}B}{\cancel{3.9}}$$

The 3.9 values cancel on the right leaving:

$$B = \frac{U - 10}{3.9}$$

The value of B can now be found from the value of U.

When transposing any formula, aim to get the unknown value on its own. Whatever operations are required on one side to achieve this, apply to the other side.

❏ *Example 1.3*

Rearrange the formula below for E to find an expression for v.

$$E = \frac{1}{2} m(v^2 - u^2)$$

Multiply both sides by 2 to get rid of the half on the right:

$$2 \times E = 2 \times \frac{1}{\cancel{2}} m(v^2 - u^2)$$

$$2E = m(v^2 - u^2)$$

Divide both sides by m:

$$\frac{2E}{m} = \frac{\cancel{m}(v^2 - u^2)}{\cancel{m}}$$

$$\frac{2E}{m} = v^2 - u^2$$

Add u^2 to both sides:

$$\frac{2E}{m} + u^2 = v^2 - \cancel{u^2} + \cancel{u^2}$$

$$\frac{2E}{m} + u^2 = v^2$$

Take the square root of each side:

$$\sqrt{\frac{2E}{m} + u^2} = \sqrt{v^2}$$

$$v = \sqrt{\frac{2E}{m} + u^2}$$

Trigonometry

Trigonometry is a widely used technique in many different types of mathematical problems. The only trigonometry required in this book is that used for calculating the relationship between different sides of a right angled triangle and one angle. The sides of a right-angled triangle are named as shown in Figure 1.1.

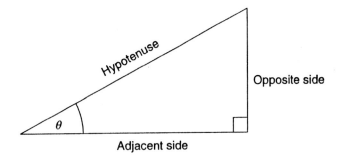

Figure 1.1

- The side opposite the right angle is called the **hypotenuse** abbreviated to 'hyp'.
- The side between the selected angle, θ, and the right angle is called the **adjacent** side, abbreviated to 'adj'.
- The side opposite the selected angle, θ, is called the **opposite side**, abbreviated to 'opp'.

The three functions of interest to us are called:

- the tangent of the angle, written as tan θ
- the cosine of the angle, written as cos θ
- the sine of the angle, written as sin θ.

We can define them using ratios of the three sides of the triangle as follows:

$$\tan \theta = \frac{\text{opp}}{\text{adj}}$$

$$\cos \theta = \frac{\text{adj}}{\text{hyp}}$$

$$\sin \theta = \frac{\text{opp}}{\text{hyp}}$$

As we are dealing with triangles only then all the angles considered are less than 90°. The tan, cos, or sin of any angle can be found from any scientific calculator. Press the button marked (DRG) to set the calculator to 'd' to measure degrees. Enter the angle, e.g. 45°, and press the required function button, e.g. (sin), to obtain the sine of 45°. To do the reverse, press the second function button, e.g. (2nd), and then the sine button (sin) again. This is usually written as $\sin^{-1}$ in text (see the following example). Sometimes there is a separate button on calculators for this reverse process marked ($\sin^{-1}$). The same method applies to cosines and tangents.

❏ *Example 1.4*

Look at Figure 1.2. Find the length of the other side shown and the angle, θ.

The hypotenuse is the side of length 4.5 m, as this is the one opposite the right angle. The opposite side is the side of length 3 m, as this is the one opposite the

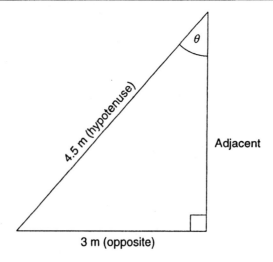

Figure 1.2

unknown angle. The adjacent side is the unknown side, as this is the side between the unknown angle and the right angle. So, what do we know?

$$\tan \theta = \frac{\text{opp}}{\text{adj}}$$

Since we do not know the angle, θ, or the adjacent side, we cannot use this one.

$$\cos \theta = \frac{\text{adj}}{\text{hyp}}$$

Again, we do not know the angle, θ, or the adjacent side.

$$\sin \theta = \frac{\text{opp}}{\text{hyp}}$$

We know both the opposite side and the hypotenuse so we can use this one to find the angle, θ.

As $\sin \theta = \dfrac{\text{opp}}{\text{hyp}}$

then $\theta = \sin^{-1}\left(\dfrac{\text{opp}}{\text{hyp}}\right) = \sin^{-1}\left(\dfrac{3\,[\text{m}]}{4.5\,[\text{m}]}\right) = 41.8°$

We can now use either the tan or the cos of the angle, θ, to find the other side.

$$\tan \theta \qquad = \frac{\text{opp}}{\text{adj}}$$

Therefore $\text{adj} = \dfrac{\text{opp}}{\tan \theta} = \dfrac{3\,[\text{m}]}{\tan 41.8°} = \dfrac{3\,[\text{m}]}{0.894} = \mathbf{3.356\,[\text{m}]}$

If we divide the sine of an angle by the cosine of the same angle the result is the tangent of that angle:

$$\frac{\sin \theta}{\cos \theta} = \frac{\text{opp}}{\text{hyp}} \div \frac{\text{adj}}{\text{hyp}}$$

$$= \frac{\text{opp}}{\text{hyp}} \times \frac{\text{hyp}}{\text{adj}}$$

$$= \frac{\text{opp}}{\text{adj}} = \tan \theta$$

$$\tan \theta = \frac{\sin \theta}{\cos \theta}$$

Consider the problem above involving an angle of 41.8°:

$$\tan 41.8° = 0.894$$

$$\cos 41.8° = 0.745$$

$$\sin 41.8° = 0.667$$

$$\tan \theta = \frac{\sin \theta}{\cos \theta} = \frac{0.667}{0.745} = 0.894$$

Theorem of Pythagoras

Pythagoras' theorem states that: the square of the hypotenuse of a right-angled triangle is equal to the sum of the squares of the other two sides. For the triangle in Figure 1.3:

$$c^2 = a^2 + b^2$$

This can be used in place of trigonometry for some problems for any right angled triangle.

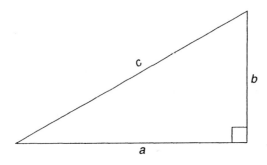

Figure 1.3

□ *Example 1.5*

The hypotenuse of a right-angled triangle is 5 m long. One of the other sides is 3 m long. Find the length of the other side.
   If we call the other side x, then:

$$5^2 = 3^2 + x^2$$

Therefore $x = \sqrt{5^2 - 3^2} = \sqrt{25 - 9}$

$$= 4$$

Differentiation

One other subject that we must touch on is calculus, specifically differentiation. There is no differentiation or integration required in this book, but sometimes quantities are expressed in differential terms, so we will briefly look at this.
   Calculus refers to a set of techniques for calculating quantities that vary rather than being fixed. In engineering, the change of a quantity with respect to time often needs to be dealt with, i.e. the rate of change. This is most easily explained with a graph (Figure 1.4).

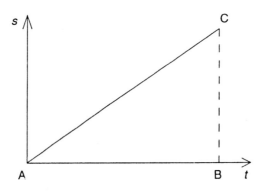

Figure 1.4

The graph shows the distance travelled, S, by a car over a period of time, t. The velocity of the car is constant. The magnitude of the velocity, v, can be calculated from:

$$\text{velocity, } v = \frac{\text{distance travelled}}{\text{time taken}} = \frac{S}{t}$$

From the graph this is also equal to the slope of the line. If the line is the hypotenuse of a right angled triangle, ABC, then:

$$\text{velocity, } v = \frac{S}{t} = \frac{\text{BC}}{\text{AB}}$$

In this situation the velocity has a constant value during the journey. Now a different journey, such as that shown in Figure 1.5, would have a different result. Here the car does not travel at a constant velocity. The velocity changes with time and so the question 'what is the velocity of the car?' cannot be answered with one value. The

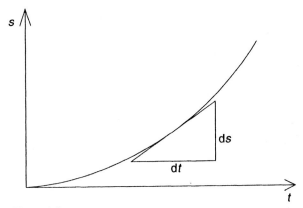

Figure 1.5

gradient of the curve changes with time. We can calculate the gradient of the curve at any one time though by drawing a tangent to the curve and finding the slope as before. If the tangent to the slope forms part of a right-angled triangle as before, then the slope of the line can be expressed in terms of the other two sides of the triangle. Instead of BC we could write dS. The d means 'a small change in'. So dS means a small change in S. Instead of AB we could write dt meaning a small change in time. As the changes in distance, S, are not constant with respect to time then we assume that if the distance travelled is considered over a very tiny time period then it would be constant during that period. At the point in time represented by where the tangent touches the curve, we can say that:

$$\text{velocity, } v = \frac{\text{distance travelled}}{\text{time taken}} = \frac{\mathrm{d}S}{\mathrm{d}t}$$

The point where the tangent touches the curve could correspond to time t. This does not seem very precise: e.g. is it at t or at $t + \mathrm{d}t$. As dt and dS are considered to be very small and approaching zero then we can say that the velocity, $v = \mathrm{d}S/\mathrm{d}t$ at a time of t. In engineering equations, instead of using a v for velocity, when the velocity may vary with time, d$S/$dt may be written to mean the velocity at any particular instant in time.

Consider now the motion of the car represented by the graph in Figure 1.6. This

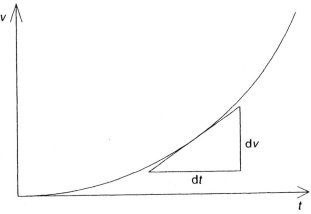

Figure 1.6

shows the velocity over a period of time. Here the velocity is changing, but not at a constant rate. For velocity changing at a constant rate, the acceleration a could be calculated by dividing the change in velocity by the time taken.

Acceleration, $a = \dfrac{v}{t}$

The same principle used above can be applied to the acceleration that varies with time.

Instantaneous acceleration a could be represented by:

$$a = \frac{dv}{dt}$$

dv/dt is then the acceleration occurring at an instant in time. dv is a small change in velocity occurring in the small time dt. We have already stated above that $v = dS/dt$. To express the acceleration in terms of the distance S, we can write:

$$a = \frac{dv}{dt} = \frac{d^2S}{dt^2}$$

$\dfrac{dS}{dt}$ is called a first differential coefficient.

$\dfrac{d^2S}{dt^2}$ is called a second differential coefficient.

In engineering, a quantity can vary in a complex manner relative to another quantity. An equation involving the motion of a body with respect to time could involve several differentials of different order. For example:

$$F = A.\frac{d^2S}{dt^2} + B.\frac{dS}{dt} + C.S$$

where A, B, and C are constants. Equations like this are called differential equations. Differential equations are used to represent dynamic relationships, i.e. with quantities that change.

In this book we will not cover how to calculate these differential equations. If you have never studied calculus before you will at least be aware of these equations' significance if they are presented in theories. However, no theories or explanations in this book rely on an understanding of calculus.

Here are some more problems. Make sure you can solve these before continuing.

Problems 1.2

11. If $G = 20$ and $y = 0.053$ then find Q, when $Q = (7.9G)^{-y} + 0.3$.

12. If $P = 17$ and $a = 0.87$ find X, when $X = \dfrac{2.8P^{0.9}}{0.6^a}$.

13. If $a = 3$, $b = 24.7$ and the angle $k = 25°$, then find P when $P = \left(\dfrac{a}{\sin k}\right)^b$.

14. If $v^2 = u^2 + 2aS$ then find an expression for S by rearranging the formula.

15. If $v = \sqrt{\dfrac{S.g.r}{2.h}}$ then find an expression for r by rearranging the formula.

16. The hypotenuse of a right-angled triangle is 3.5 m and one of the other sides is 2.5 m. Using the theorem of Pythagoras calculate the length of the remaining side.

17. The hypotenuse of a right-angled triangle is 9.6 m and one of the other sides is 5 m. Find the two unknown angles.

18. The hypotenuse of a right-angled triangle is 0.8 m. One of the angles is 25°. Find the length of the other two sides and the other unknown angle.

19. One angle of a right-angled triangle is 55°. What is the other unknown angle?

20. Find y when $x = 14$, if $y = 2x^2 + 4x + 6$.

2 Forces

2.1 Mass, volume and density

This chapter deals primarily with mass. Mass is one of the fundamental quantities. It is important to understand the effect that the Earth has on a particular quantity of mass and also to be able to work out how much room this mass takes up. The quantities of weight, volume and density are also introduced in this chapter, as they are all related.

Mass

The mass of an object is a measure of the quantity of matter that makes up that object. This quantity of matter will remain the same wherever and however the object is positioned. We can squash it, stretch it or do anything to it so long as we do not chop bits off it: the mass will not change. Therefore the mass of an object is a constant property, i.e. it will always remain the same.

The unit of mass is the kilogram and the unit symbol is kg. The quantity symbol is m. So if data is given about a piston for example, and $m = 0.5$ kg, you know that this refers to the mass. Another common unit for mass is the tonne which equals 1000 kg. The kilogram is not the unit of weight and it is not the unit of force. However the units of weight and force are derived from the unit of mass.

Force

If we push an object, it tends to move. This 'push' is an example of a force. A force is difficult to imagine or define. A common definition is based on its effects. The unit of force is called the newton and the unit symbol is N.

1 N is the force required to give a mass of 1 kg an acceleration of 1 m/s^2, that is an increase in velocity of one metre per second in each second (see Figure 2.1).

Expressed mathematically,

$$1 \text{ N} = 1 \text{ kg} \times 1 \text{ m/s}^2$$

This relationship between force, mass and acceleration was first suggested by Sir Isaac Newton and is part of his second law of motion. The quantity symbol for force is F. It is normally written as follows:

$$F = m \times a \qquad (a = \text{acceleration})$$

Newton's first law of motion suggests a force as: that action which changes or tends to change the motion of the body on which it acts. A simpler definition could be: the

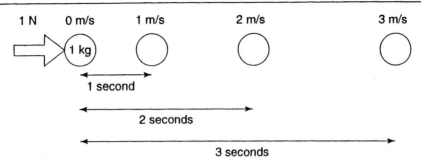

1 N   0 m/s   1 m/s   2 m/s   3 m/s

1 kg

1 second

2 seconds

3 seconds

Figure 2.1

action of one body on another. It is Newton's second law, remember, that enables us to measure a force: $F = m \times a$. The term 'body' is commonly used in science. This just means a mass or object that is being considered.

Weight

Newton also developed a law which states: that there is an attractive force between all bodies; the magnitude of the force depends upon the mass of each body and their distance apart. For two cars that are on the surface of the Earth, the force between them is very small. However, the attractive force between one of the cars and the Earth is much larger as the mass of the Earth is so large. This force is what we usually call weight and it has the same units as force, the newton [N].

Newton's second law of motion ($F = m.a$) explains that for a mass close to the Earth's surface there is an acceleration. This acceleration is due to gravity, which is the Earth's force of attraction, and is represented by the symbol g. For bodies falling to the Earth the acceleration then is g. The value of g varies slightly over the surface of the Earth but the value generally used for engineering is 9.81 m/s². The acceleration is always towards the centre of the Earth (see Figure 2.2).

The idea of mass is often confused with that of weight which can be dangerous in engineering calculations and causes no end of problems. If you asked someone the weight of an engine because you wanted to lift it out of a vehicle, the answer they

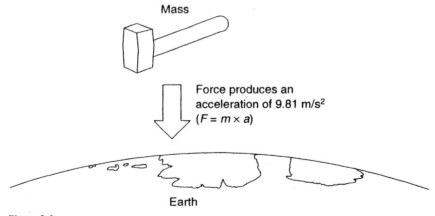

Mass

Force produces an acceleration of 9.81 m/s² ($F = m \times a$)

Earth

Figure 2.2

would probably give would be a mass and not a weight; for example 100 kg. This would probably not cause any confusion because anywhere on Earth 100 kg has a mass of approximately the same weight. If you were to transport that engine to the Moon, the mass would remain the same but the weight would change, because the Moon has a different mass to that of the Earth. So the attractive force and the acceleration due to the Moon's gravity are different. The main point of this is that although the mass remains the same, the weight is not a fixed quantity. If the engine were then transported into outer space, the mass would still remain at 100 kg but the weight would be zero, since there would be no gravitational force at all.

To find out the weight of something, remember that the weight is the attractive force due to the Earth's acceleration and use the formula $F = m \times a$. It might seem a bit strange talking about acceleration when something could be stationary. With the Earth the acceleration is g which equals 9.81 m/s^2, so:

$$\text{gravitational force} = m \times a = m \times g$$
$$\text{The weight of the engine} = 100\,[\text{kg}] \times 9.81\,[\text{m/s}^2]$$
$$= 981\text{ kg m/s}^2 = 981\text{ N}$$
$$\text{Notice that } 1\text{ kg m/s}^2 = 1\text{ N}.$$

❏ *Example 2.1*

The Moon buggy that was taken on the Apollo Moon landing in July 1971 had a mass of 699 kg. What did this weigh on Earth and what did it weigh on the Moon? On the Moon the acceleration due to gravity, $g_{\text{moon}} = 1.624$ m/s^2.

$$\text{Earth's gravitational force (weight)} = m \times g_{\text{earth}}$$
$$= 699\,[\text{kg}] \times 9.81\,[\text{m/s}^2]$$
$$= 6857.19\text{ kg m/s}^2$$
$$= \mathbf{6857.19\text{ N}}$$

$$\text{Moon's gravitational force} = 699\,[\text{kg}] \times 1.624\,[\text{m/s}^2]$$
$$= 1135.18\text{ kg m/s}^2$$
$$= \mathbf{1135.18\text{ N}}$$

What would the mass of the buggy be in outer space whilst being transported to the Moon and what would its weight be? Answer: Its mass would be the 699 kg, as mass is a constant property. The weight however would be zero as there is no gravitational force.

Area

Area is the measurement of the size of a surface. A linear dimension is a length that can be measured in a straight line. An area is said to be a measurement in two dimensions. It is measured as the product of two unit lengths; in the metric system the unit of length is the metre and so the measure of area is metre × metre which is known as square metres. The area of a surface in the metric system is the number of square metres that will fit into this surface (Figure 2.3).

The SI unit symbol for the square metre is m^2 and the quantity symbol for area is A.

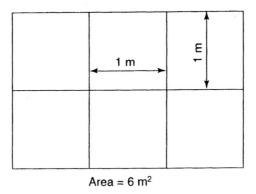

Area = 6 m²

Figure 2.3

It is important to know the formula for the area of certain common shapes. Most of them you will probably be familiar with. These are shown as follows:

The **square** or **rectangle**: Area = length of one side × length of the other side. This is usually referred to as length × breadth (Figure 2.4).

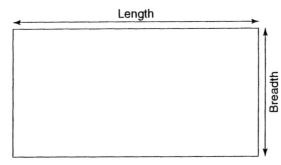

Figure 2.4

The **triangle** can be positioned any way up for this calculation. Measure the side of the triangle at the bottom and then measure the vertical height. If one of the angles is a right angle then the vertical height could also be the length of one of the sides. The area of the triangle is the base multiplied by the vertical height and then divided by two (Figure 2.5).

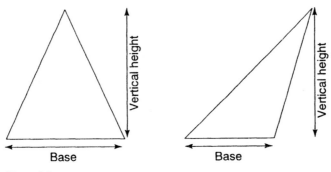

Figure 2.5

$$\text{Area} = \frac{\text{base} \times \text{vertical height}}{2}$$

The **circle**: To measure the area of a circle we need to know the radius. The area is then equal to 3.142 × radius × radius. The number 3.142 is represented by the Greek letter π pronounced pi. A more accurate value of pi would be 3.14159265 but 3.142 is accurate enough for most engineering calculations. Most calculators have a pre-set value available by pressing a separate button which is more convenient than having to enter the number each time (Figure 2.6).

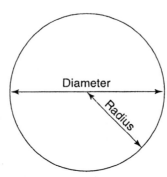

Figure 2.6

$A = \pi \times r^2 \qquad (r = \text{radius})$

This is also equal to $\dfrac{\pi d^2}{4} \qquad (d = \text{diameter})$

If you wanted to find the area of a flat annular ring (doughnut shape) then find out the area of the larger circle and then the area of the smaller circle and subtract the smaller from the larger (Figure 2.7).

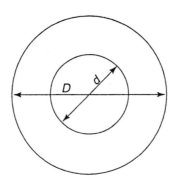

Figure 2.7

$$\text{Area} = \frac{\pi D^2}{4} - \frac{\pi d^2}{4} = \frac{\pi}{4} \times (D^2 - d^2)$$

❑ *Example 2.2*

Find the cross-sectional area of the cylinder liner shown in Figure 2.8. To find the cross-sectional area of an illustrated component, just imagine that it is cut in two where indicated and consider the chopped surface. Cross-sectional area is normally abbreviated to CSA and shown shaded on drawings.

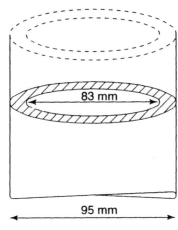

Figure 2.8

$$\text{Area} = \frac{\pi}{4} \times (D^2 - d^2)$$

$$= \frac{\pi}{4} \times ((95 \times 10^{-3})^2 - (83 \times 10^{-3})^2) \,[\text{m}^2]$$

$$= \frac{\pi}{4} \times (95^2 - 83^2) \times 10^{-6} \,[\text{m}^2]$$

$$= 1677.610 \times 10^{-6} \,\text{m}^2$$

At this stage we could convert the answer to mm² if we wanted but first we need to find out how many mm² there are in 1 m².

$$1000 \text{ mm} = 1 \text{ m}$$
$$(1000 \text{ mm})^2 = (1 \text{ m})^2$$
$$10^6 \text{ mm}^2 = 1 \text{ m}^2$$
$$1 = 10^6 \left[\frac{\text{mm}^2}{\text{m}^2}\right] = 10^{-6} \left[\frac{\text{m}^2}{\text{mm}^2}\right]$$

Note we have now produced a unity bracket in both its forms for converting mm² to m² and m² to mm².

$$\text{Area} = 1677.610 \times 10^{-6} \,[\cancel{\text{m}^2}] \times 10^6 \left[\frac{\text{mm}^2}{\cancel{\text{m}^2}}\right]$$

$$= 1677.610 \,\text{mm}^2$$

Alternatively we could have worked through the entire question in mm². Don't be tempted to miss out the units at any stage – you will probably go wrong.

If you wanted to find the length of the circumference of a circle then use the formula $\pi \times d$. This is sometimes useful for finding the area of the surface of a cylinder. It is easier when finding the area of curved surfaces if you imagine them to be flattened out first (Figure 2.9).

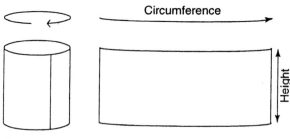

Figure 2.9

Once flattened out the surface of the cylinder can then be calculated as a rectangle, i.e. length × breadth.

$$\text{Area} = \pi d \times h \qquad (h = \text{height})$$

☐ *Example 2.3*

Find the area of the wall of an engine cylinder that is swept by a piston ring. The bore is 83 mm and the stroke is 73 mm.

$$\begin{aligned}
\text{Area} &= \pi d \times h \\
&= \pi \times 83 \text{ mm} \times 73 \text{ mm} \\
&= 19.034 \times 10^3 \text{ mm}^2
\end{aligned}$$

We can convert this to m² using the unity bracket from before.

$$\text{Area} = 19.034 \times 10^3 \, [\cancel{\text{mm}^2}] \times 10^{-6} \left[\frac{\text{m}^2}{\cancel{\text{mm}^2}} \right]$$

$$= 19.034 \times 10^{-3} \text{ m}^2$$

A **trapezium** is a four-sided shape with only two sides that are parallel (Figure 2.10).

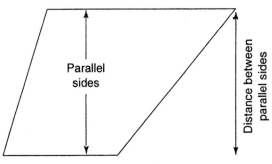

Figure 2.10

$$\text{Area} = \frac{\text{sum of the parallel sides}}{2} \times \text{distance between them}$$

Volume

The volume of an object is a measure of how much space it will fill. Volume is measured in three dimensions and the unit is the product of three unit lengths. In the metric system, since the unit of length is the metre then the unit of volume is the cubic metre. The unit symbol is m^3 ($[m] \times [m] \times [m] = [m^3]$) and the quantity symbol is V.

The volume of a tank for instance, is the number of cubes with a side length of one metre that will fit into that tank. If the tank is rectangular in shape then the volume is the product of the straight sides.

☐ Example 2.4

Find the volume of a petrol tank that measured 0.4 m by 0.6 m by 0.2 m (see Figure 2.11).

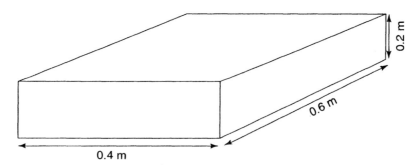

Figure 2.11

$$V = 0.4\,[m] \times 0.6\,[m] \times 0.2\,[m] = \textbf{0.048 m}^3$$

As $1\,m^3 = 1000$ litres then the unity bracket is:

$$1 = \frac{1000}{1}\left[\frac{\text{litres}}{m^3}\right] = \frac{1}{1000}\left[\frac{m^3}{\text{litres}}\right].$$

$$0.048\,[\cancel{m^3}] \times 1000 \left[\frac{\text{litres}}{\cancel{m^3}}\right] = \textbf{48 litres}$$

Remember we use the form of the unity bracket so that the unit we want to get rid of (cancel out) is on the bottom.

If the sides of the tank are not all straight then look for a way that it could be positioned on a level surface so that all the sides are straight and vertical. The volume then equals the area of the base multiplied by the height (Figure 2.12).

The volume of this tank is as follows:

base = πr^2
volume = $\pi r^2 \times h$ where h = height

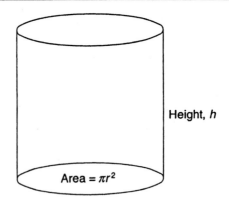

Figure 2.12

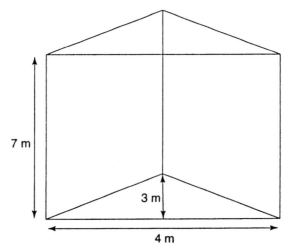

7 m

3 m

4 m

Figure 2.13

The volume of this tank is also base area × vertical height:

$$\text{Volume} = \frac{3 \times 4}{2}\,[\text{m}^2] \times 7\,[\text{m}] = 42\ \text{m}^3$$

(See Figure 2.13).

☐ *Example 2.5*

Find the swept volume of the piston in Example 2.3. To find the swept volume of something, imagine the piston of an engine at the bottom of its stroke, i.e. bottom dead centre, and consider the volume in the cylinder above the piston. Now imagine the piston at the top of its stroke, i.e. top dead centre. The volume above the piston now is called the clearance volume. The swept volume is the maximum volume minus the clearance volume, i.e. the cylinder volume that the piston sweeps through. The sum of all the swept volumes of an engine refers to the capacity of the engine. The compression ratio of an engine is the ratio of the maximum volume to the clearance volume (see Figure 2.14).

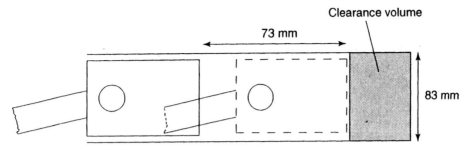

Figure 2.14

Swept volume = $\pi r^2 \times$ length of stroke

$$= \pi \times \frac{(83)^2}{4} \times 73 \,[\text{mm}^2].[\text{mm}]$$

$$= \textbf{394.9744} \times \textbf{10}^3 \,\textbf{mm}^3$$

If this is a four cylinder engine then the total capacity of the engine is :

$$4 \times 394.9744 \times 10^3 \,\text{mm}^3 = 1579.8976 \times 10^3 \,\text{mm}^3$$

Most car engine capacities are usually expressed in litres so we could convert mm^3 to litres.

$$1 \,\text{m}^3 = 10^9 \,\text{mm}^3$$

so the unity bracket $= 10^9 \left[\dfrac{\text{mm}^3}{\text{m}^3}\right] = 10^{-9} \left[\dfrac{\text{m}^3}{\text{mm}^3}\right]$

Also $1 \,\text{m}^3 = 1000$ litres

so the unity bracket $= 10^3 \left[\dfrac{\text{litres}}{\text{m}^3}\right] = 10^{-3} \left[\dfrac{\text{m}^3}{\text{litres}}\right]$

So engine capacity $= 1579.8976 \times 10^3 \,[\text{mm}^3] \times 10^{-9} \left[\dfrac{\text{m}^3}{\text{mm}^3}\right] \times 10^3 \left[\dfrac{\text{litres}}{\text{m}^3}\right]$

$$= 1579.8976 \times 10^{-3} \,\text{litres} = 1.58 \,\text{litres}$$

This would be classed as a 1.6 litre engine.

If the clearance volume is $47.0768 \times 10^3 \,\text{mm}^3$ then what is the compression ratio?

Compression ratio $= \dfrac{\text{swept volume} + \text{clearance volume}}{\text{clearance volume}}$

$$= \frac{(394.9744 \times 10^3) + (47.0768 \times 10^3)}{(47.0768 \times 10^3)} \left[\frac{\text{mm}^3}{\text{mm}^3}\right]$$

We can see at this stage that the units of the calculation cancel out, meaning that there are no units of compression ratio. Also the 10^3 terms cancel out on the top and bottom. This leaves us with:

Compression ratio $= \dfrac{394.9744 + 47.0768}{47.0768} = \textbf{9.39}$

This would normally be written as 9.39:1. This tells you how many times bigger the maximum volume is than the clearance volume.

Density

Equal masses of different materials occupy different volumes. Consider 1 kg of lead and 1 kg of feathers: they have the same mass but their volumes are very different. Density is a measure of the ability of a material to pack more or less into a given volume.

Density is the mass of a unit volume of a material, so

$$\text{density} = \frac{\text{mass of material}}{\text{volume occupied}}$$

The quantity symbol is ρ which is a Greek symbol pronounced 'ro'. The units of density will be mass unit per volume units and are measured in kg/m^3 although you may come across other units within the metric system, e.g. $tonne/m^3$.

The density of a substance depends on volume. Volume varies with temperature, so when a density is stated the temperature at which it was determined should also be stated. With solids and liquids the changes in density are fairly small over normal motor vehicle operating temperatures, but the density of gas varies greatly, even with quite small changes in temperature. Intercoolers are often fitted to cool an engine's inducted air after a turbo charger has compressed and heated it a few degrees Celsius in order to reduce the density. This can almost account for as much benefit as the turbo charger itself.

Relative density

Relative density is sometimes quoted as it allows the comparison between the density of one substance and another:

$$\text{Relative density} = \frac{\text{density}}{\text{density of fresh water}} \quad \text{(in the same units)}$$

There are no units of relative density as they cancel out. The density of fresh water is $1000\ kg/m^3$ and its relative density must be one. The density of mild steel is $7860\ kg/m^3$ and so its relative density is 7.86. In the past, relative density was called *specific gravity*, a term you may still come across.

❑ Example 2.6

How much does the weight of a van increase by if it is filled up with 75 litres of fuel? The relative density of the fuel at the filling temperature is 0.9.

Density = relative density × density of fresh water
$$= 0.9 \times 1000\ [kg/m^3]$$
$$= 900\ [kg/m^3]$$

As $\rho = \dfrac{m}{V}$ then $m = \rho \times V$

$$m = 900 \left[\frac{kg}{m^3}\right] \times 75 \, [\text{litres}] \times 10^{-3} \left[\frac{m^3}{\text{litres}}\right]$$

$$= 67.5 \, kg$$

$$\text{Weight} = m \times g = 67.5 \, [\text{kg}] \times 9.81 \, [\text{m/s}^2]$$
$$= 622.175 \, \text{kg m/s}^2 = \mathbf{622.175 \, N}$$

2.2 Forces

Introduction

In Section 2.1 on mass we looked at the definition of force and its relationship with mass and acceleration. In this Section we look at forces in more detail and see how they act on a body and what happens when several forces act together.

A thorough understanding of force is very important. The whole subject of mechanics is based on how forces act on matter, whether it is stationary or in motion. In a motor vehicle, a force of some sort acts on every part. The force of gravity acts on the vehicle, affecting the design of all components. Power is transferred from the combustion space of the engine through the transmission system to the wheels of the vehicle, all because of forces (Figure 2.15).

The gas pressure causes a force to act down on the piston crown. This 'push' is transferred to the connecting rod and to the crankshaft, converting linear motion into rotary motion.

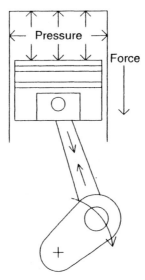

Figure 2.15

Force as a vector quantity

Quantities dealt with in engineering problems can usually be put in one of two groups: scalar quantities and vector quantities. A scalar quantity can be completely described by its magnitude and the units. Examples are length, mass and temperature. Scalar quantities that are measured in the same units can be dealt with by simple arithmetic, as we have seen with multiplication, division, addition and subtraction on mass calculations. To describe a vector quantity, the point of application and the direction need to be specified as well as the magnitude and units. Examples are force, velocity and acceleration. A vector can best be represented graphically as shown in Figures 2.16 and 2.17.

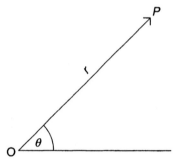

Figure 2.16 Polar co-ordinates

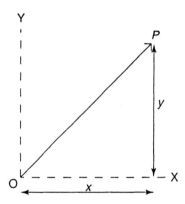

Figure 2.17 Rectangular co-ordinates

These are called vector diagrams. The direction of the arrows is in the same direction as the forces. The length of the line is proportional to the magnitude of the force. For instance if there was a force of magnitude 10 N and you decided to have a scale of 10 mm = 1 N, then the line representing this would be 100 mm long. Look at the polar co-ordinate diagram: when measuring the angle of a vector it is usual to assume that the three o'clock position is zero degrees and angles are measured anticlockwise. The vector OP can be described by the length, r, and the angle, θ, as $r\theta$. Now, the rectangular co-ordinate diagram: the same vector, OP, can be described by the two lengths, x and y, as x,y.

Vector quantities can be multiplied by scalar quantities. The result will be a vector quantity with direction the same as the initial vector quantity. If we consider a vector

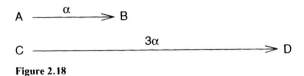

Figure 2.18

with a magnitude of α and then multiply it by 3 for instance, the result will be a vector in the same direction but with a magnitude of 3α. See Figure 2.18.

If AB = α then 3α can be represented by CD, which is three times the length of AB, but in the same direction.

Vector addition

Two vectors can be added up using the parallelogram rule. In this, the two vectors, A and B, are drawn starting from the same point, O. The magnitude of the vectors is represented by the length of the lines. A parallelogram is then drawn using the vectors as the adjacent sides. The diagonal of the parallelogram, C, drawn from the origin then represents the sum or the resultant of the two forces both in magnitude and direction (Figure 2.19).

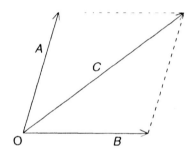

Figure 2.19 Parallelogram rule

The parallelogram law shows an important feature of vectors: it does not matter in which order vectors are added. Two vectors can also be added using the triangular rule or 'head to tail' rule. The two vectors, A and B, are drawn head to tail in any order and the resultant vector, C, or their sum is drawn from the tail of the first one to the head of the second one. A *resultant* vector is one that has the same effect as several other forces acting together through the same point. The term 'concurrent forces' is used to describe forces that act through the same point (Figure 2.20).

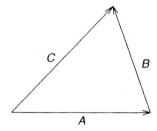

Figure 2.20 Triangular rule or 'head to tail' rule

In fact the head to tail rule can be applied to any number of vectors added in any order. The resultant vector is drawn from the tail of the first one to the head of the last one. This type of diagram is known as a force polygon. A polygon is a closed shape with a number of straight sides; e.g. a triangle is a three-sided polygon and a hexagon is a six-sided polygon (Figure 2.21).

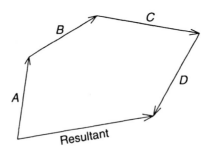

Figure 2.21 Force polygon

The same rules apply if vectors are added that act in the same direction (Figure 2.22).

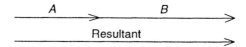

Figure 2.22 Addition of vectors acting in the same direction

To subtract a vector then reverse the direction (Figure 2.23).

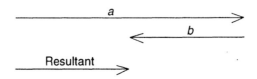

Figure 2.23 Subtraction of vectors acting in the same direction

❏ *Example 2.7*

Two forces **A** and **B** are shown. Draw the vectors of **A** + **B** and **A** − **B**.

In the example in Figure 2.24, the magnitude of the vectors is unknown but the head

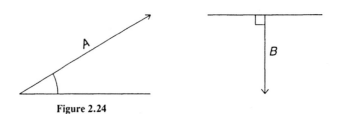

Figure 2.24

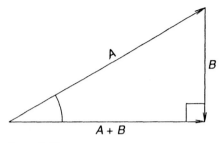

Figure 2.25

to tail rule can be demonstrated. Draw the vectors **A** and **B**, applying the head to tail rule (Figure 2.25).

The length of the vector **A** + **B** is proportional to its magnitude. The length of this would depend on the magnitude of **A** and **B** and the scale of your diagram. To find the vector **A** − **B**, vector **B** needs to be subtracted from **A**. To subtract the vector **B**, it goes in the opposite direction and the resultant is found in the same way (Figure 2.26).

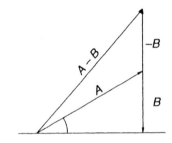

Figure 2.26

The effects of a force

To understand that the point of application and the direction are important when describing a force, think about the small truck shown in Figures 2.27 and 2.28. In each case a force of the same magnitude is exerted but not always with the same point of application and direction.

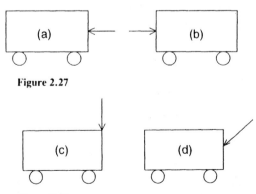

Figure 2.27

Figure 2.28

- Truck (a) would move to the left.
- Truck (b) would move to the right.
- The left end of truck (c) would move up and the right end would move down.
- The right end of truck (d) would move down and the truck would tend to move to the left.

None of the four forces shown is the same. Although they are of the same size, the ·effect of each on the truck is clearly different. This shows that force is a vector quantity. Because different forces have different effects, it is important to fully describe each force by its magnitude, units, point of application and direction.

Space diagrams

A space diagram is an illustration of a system of vectors drawn to simplify the situation. You will often see an exam type problem presented with a space diagram to show the points of application and direction of forces acting on a body: If one is not presented with the question, then you should draw one. The diagram usually does not show objects that are in contact with the body, such as supports, but just the forces that these exert on the body. If a system of forces has exact points of application on a body then it may be necessary to draw the space diagram to some convenient scale. For a stationary steel block resting on the surface of the Earth the space diagram could look like that in Figure 2.29.

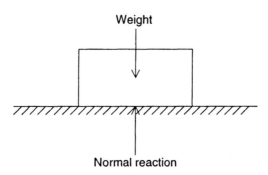

Figure 2.29

The downward force is the weight of the block acting through the centre of gravity. The upward force is the normal reaction of the surface of the Earth. The normal reaction may seem a strange idea at first because we are not usually aware of it. For every action there is an equal and opposite reaction – more of this later. If the block is stationary the forces acting on it must be in equilibrium and the downward force caused by gravity is balanced by an upward force from the Earth. The space diagrams are important for showing the situation in an uncluttered form.

Resolution of forces

It is sometimes useful to think of a single force acting on a point as being split into other force components. This is really the reverse of adding forces to find one resultant force. Usually two rectangular force components are the most useful and

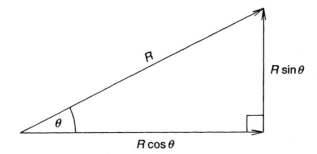

Figure 2.30 Resolution of a force *R*

also the easiest to work out as we are dealing with a right-angled triangle. Look at the example in Figure 2.30.

The force **R** can be split in to two rectangular force components, the vertical component and the horizontal component.

$$\sin \sigma = \frac{\text{opposite side}}{\text{hypotenuse}} = \frac{\text{vertical component}}{\mathbf{R}}$$

Therefore vertical component = **R** sin σ

$$\cos \sigma = \frac{\text{adjacent side}}{\text{hypotenuse}} = \frac{\text{horizontal component}}{\mathbf{R}}$$

Therefore horizontal component = **R** cos σ

After a few examples you will get used to knowing that the side adjacent to the angle is **R** cos σ and the side opposite the angle is **R** sin σ.

Example 2.8

A force of 20 N act at an angle of 35° to the horizontal plane. What are the vertical and horizontal components?

Horizontal component = 20 [N] × cos 35° = 16.383 N
Vertical component = 20 [N] × sin 35° = 11.472 N

Forces in equilibrium

A system of forces acting on a body is said to be in *equilibrium* if there is no tendency for the body on which they act to move. This just means that, although there may be several forces acting on something, their effects cancel each other out. If you hear the term 'statics', this is the study of stationary forces in equilibrium. (A state of equilibrium may also exist when a body is in a state of uniform motion.) The term 'kinetics' on the other hand is the study of forces which cause motion. The *equilibrant* of a concurrent force system (forces that act through the same point remember) is a force that, when applied to the system, would place it in a state of equilibrium. The equilibrant must therefore be a force that has the same magnitude as the resultant but acts in the opposite direction. It is probably easier to show this with an example. Look at the force polygon example above again, as shown in Figure 2.31.

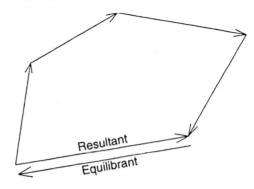

Figure 2.31

The resultant force is the force that produces the same effect as all the other forces acting together. The equilibrant is a force that acts along the same line of action as the resultant, but in the opposite direction. If this equilibrant were applied it would bring the system into a state of equilibrium. A way of defining a system of concurrent forces in equilibrium is that the force polygon closes.

When a system consists of only three forces that are co-planar, if it is to be in equilibrium, the three forces must pass through the same point, i.e. they must be concurrent. *Co-planar* means that all the forces act in one plane. If we represent the forces by a closed polygon, it would in this case be a triangle. When the forces do not act at one point then turning effects are introduced. We will look at turning effects later, but just remember for now that when several forces all act through the same point there are no turning effects. It is not necessary for forces acting on a body to all pass through the same point for equilibrium, when more than three forces are present. It is just necessary for all components in a given direction to equal zero, e.g. all the $R \sin \sigma$ and $R \cos \sigma$ components (resolution of forces) added together must equal zero. Example 2.9 involves triangles of forces.

☐ *Example 2.9*

A man weighing 80 kg stands a ladder on a rough horizontal floor and leans it against a 'perfectly smooth' wall at an angle of 30° to the vertical. He then climbs half-way up the ladder. Determine graphically the direction and magnitude of the reaction between the ladder and the wall, and the floor and the ladder.

Determine graphically means that you work the problem out using diagrams rather than maths. It helps if you have a set-square and a protractor. The first thing to do is to draw a space diagram. This will help to clarify the situation. Assuming that the ladder does not slip, then we can say that this is a system in equilibrium. Also as this is a triangle of forces all the forces will pass through one point, i.e. they are concurrent. It is important to find the point of concurrency, which will not always be obvious. This can be done on the space diagram shown in Figure 2.32.

Draw the ladder at 30° to the wall. The ladder's length does not matter. Because in the question it states that the wall is perfectly smooth then you can assume that the reaction between the wall and the ladder, which we can call R_w, is normal to the wall, (i.e. sticks out at 90°). This is an typical assumption made in questions to simplify the

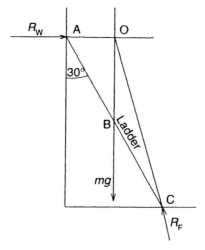

Figure 2.32

problem. So draw a horizontal line to represent this at the top of the ladder and normal to the wall. Next draw the force line of the weight of the man. This will be vertically downwards half-way down the ladder. Where these two force lines cross is the point of concurrency, O. The other force, the reaction between the floor and the ladder, must also pass through this point. The last line on the space diagram can now be drawn in from the bottom of the ladder through the point of concurrency O. From this space diagram you cannot determine any value but you now know the direction of all the forces and can draw a force diagram. Pick a convenient scale. Draw the weight of the man first.

Weight $= mg = 80\,[\mathrm{kg}] \times 9.81\,[\mathrm{m/s^2}] = 784.8\,\mathrm{N}$

Remember that $1\,\mathrm{kg\,m/s^2} = 1\,\mathrm{N}$. So you could make your scale say $20\,\mathrm{N} = 5\,\mathrm{mm}$ to make the line about 200 mm long. If you can draw the space diagram and the force diagram on the same piece of paper it makes things easier. See Figure 2.33. Draw 'fd' to scale parallel to OB, i.e. vertically, and the length proportional to 785 N. At 'd'

Figure 2.33

draw 'de' parallel to AO, i.e. horizontally. At 'f' draw 'fe' parallel to CO. Do this by sliding a set-square along a ruler from the space diagram to the force diagram. The point 'e' completes the triangle. You can now measure the angle with a protractor. It should be around 74°. If you measure the line 'ef' according to your scale it should give a value of 816 N. The direction of this is 74° to the horizontal from 'e' to 'f'. We usually assume that the 3 o'clock position is zero degrees and measure anticlockwise from there, so the actual angle of direction would be 106° (= 180° − 74°). The line 'de' should give you an answer of 225 N. The direction of this we know as being normal to the wall which is an angle of 0°.

Triangle of forces – practical applications

The **jib crane**: This is used for supporting heavy parts of machinery. In Figure 2.34, AC is the post, BC is the tie and AB is the jib. If a mass is suspended from B then there are three forces in equilibrium at B, the crane head. These are the gravitational force on the mass, mg, the pull in the tie T and the push in the jib E. The push of the jib E is the equilibrant of the pull of the tie T and the gravitational effect on the mass mg. Therefore E is the equal and opposite to the resultant R of T and mg.

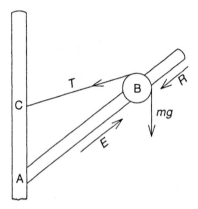

Figure 2.34 The jib crane

The **reciprocating engine** converts the reciprocating motion of the piston into the rotary motion of the crankshaft, O (Figure 2.35). Think about the forces meeting at the gudgeon pin (where the top end of the connecting rod is connected to the piston) when the cylinder is on its power stroke. The piston pushes vertically downwards. The thrust in the connecting rod appears as an upward resisting force at its top end inclined to the vertical by a varying angle α. There is a third force that balances the horizontal component of the thrust in the connecting rod that acts between the piston and the cylinder walls. As the piston force is always vertically down and the side thrust is always horizontal then the force diagram is always a right angled triangle (Figure 2.36).

This simplifies problems and makes the diagram easy to draw. The angle between the centre line of the engine and the connecting rod is shown as α in the space diagram. This is the same angle between the piston force and the force in the connecting rod on the force diagram.

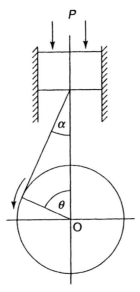

Figure 2.35 Reciprocating engine mechanisms

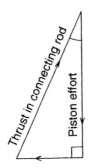

Figure 2.36

When trying the force problems at the end of the chapter, work slowly and carefully, and try to keep any drawings neat and as large as possible. Refer back to the last few paragraphs when you need to.

2.3 Moments

Introduction

A **moment** of a force is the turning effect of a force about a point. It is calculated as the product of the force and the distance from the pivot point or axis, perpendicular to the line of action (Figure 2.37).

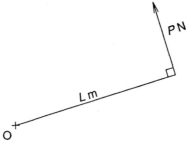

Figure 2.37

The moment of the force $P[N]$ about the point $O = P \times L$, where L is the distance from O perpendicular to the line of action. The moment of a force about a point is only zero when either the force is zero, or the line of action of the force passes through the point. To show which way around the point the force acts, we use the terms clockwise and anti-clockwise moments. In calculations we use the convention of positive moments indicating a clockwise direction and negative moments indicating an anti-clockwise direction. The unit of a moment is the product of force and length, the newton metre, [Nm]. When moments of a force produce a turning moment or a twisting moment then this is sometimes referred to as torque. This means that the units of torque are the same as for the moment of a force. This will be covered in more detail in the section on torsion later. You may be familiar with the term torque from engine overhauls or in relation to engine performance. It is important that some engine nuts are tightened to a certain torque either because the components need to be pushed together with a certain force or to prevent damage to the bolt threads. The torque that an engine produces at different revolutions will affect the clutch and gearing required to give the desired performance of the vehicle.

☐ *Example 2.10*

See Figure 2.38. Forces of 60 N, 50 N, 80 N and 120 N act along the sides of the square shown which has a side length of 2 m. Find the sum of the moments about points A and B.

$$\text{Clockwise moments about A} = (80\,[N] \times 1\,[m]) + (50\,[N] \times 1\,[m])$$
$$= 130 \text{ Nm}$$
$$\text{Anti-clockwise moments about A} = -((120\,[N] \times 1\,[m]) + (60\,[N] \times 1\,[m]))$$
$$= -180 \text{ Nm}$$

Notice we use the negative sign to indicate anti-clockwise moments.

$$\text{The resultant moments about A} = 130\,[Nm] - 180\,[Nm] = -50 \text{ Nm}$$
$$\text{Resultant moments about A} = \mathbf{-50 \text{ Nm, anti-clockwise}}$$
$$\text{Clockwise moments about B} = (80\,[N] \times 2\,[m]) + (50\,[N] \times 2\,[m])$$
$$= 260 \text{ Nm}$$
$$\text{Anti-clockwise moments about B} = \text{zero}$$
$$\text{Resultant moments about B} = \mathbf{260 \text{ Nm, clockwise}}$$

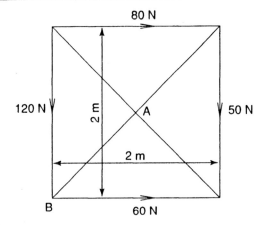

Figure 2.38

Balance of moments

If a body is subjected to a number of forces and is at rest, then there must be a balance of moments or the resultant moment would cause rotation of the body. For a state of equilibrium:

clockwise moments = anti-clockwise moments

This is called the principle of moments.

The best way of showing how this principle can be applied to solve problems is by looking at a few examples. Many problems you will come across will involve a beam of negligible mass supported on knife edges. It is helpful to think of the beam as having no mass and being supported on knife edges, to simplify the problem. The problem is then reduced to forces (including reactions from supports) acting at precise points along the beam. Many practical problems can be solved by simplifying them in this way.

☐ *Example 2.11*

The beam shown has a negligible mass and is supported on a knife edge 0.75 m from the left side. Two forces are applied, one of 50 N at the left end and the other of 20 N at a point 0.5 m to the left of the support (see Figure 2.39). The beam may pivot on

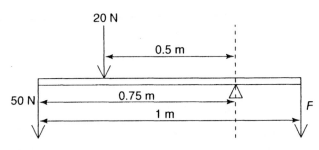

Figure 2.39

the support. What force needs to be applied at the right end to put the beam in a state of equilibrium?

To solve this we apply the principle of moments:

clockwise moments = anti-clockwise moments

If F is the unknown quantity then:

$(F\,[\text{N}] \times 0.25\,[\text{m}]) = (50\,[\text{N}] \times 0.75\,[\text{m}]) + (20\,[\text{N}] \times 0.5\,[\text{m}])$
$(F\,[\text{N}] \times 0.25\,[\text{m}]) = 47.5\,\text{Nm}$

$$F = \frac{47.5\,[\text{N}\cancel{m}]}{0.25\,[\cancel{m}]} = \textbf{190 N}$$

❑ *Example 2.12*

The beam shown in Figure 2.40 has a negligible mass and pivots around the fulcrum at the far right end. A force of 100 N acts downwards 0.5 m from the right end. What upward force applied at the far left end is required to balance the beam?

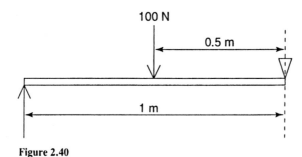

Figure 2.40

Again we apply the principle of moments:

clockwise moments = anti-clockwise moments
$(F\,[\text{N}] \times 1\,[\text{m}]) = (100\,[\text{N}] \times 0.5\,[\text{m}])$

$$F = \frac{(100\,[\text{N}] \times 0.5\,[\cancel{m}])}{1\,[\cancel{m}]} = \textbf{50 N}$$

❑ *Example 2.13*

A beam rests on two knife-edge supports as shown in Figure 2.41. Find the reaction forces that the knife-edges apply to the beam.

Consider one end of the beam as a pivot, even though the forces are in equilibrium, and apply the principle of moments. Take the left end as the pivot:

clockwise moments = anti-clockwise moments

If we represent a reaction force by R then:

$(100\,[\text{N}] \times 0.65\,[\text{m}]) = (R_2 \times 0.85\,[\text{m}]) + (R_1 \times 0.15\,[\text{m}])$
$65\,[\text{Nm}] = 0.15R_1 + 0.85R_2$ \hfill (1)

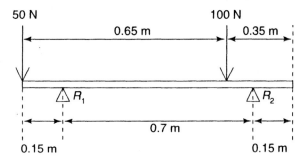

Figure 2.41

Notice that we ignore the 50 N force as it passes through the fulcrum, as the moment of the force is zero, $50 \, [\text{N}] \times 0 \, [\text{m}] = 0$

Now consider the other end of the beam as a pivot and apply the principle of moments:

clockwise moments = anti-clockwise moments

$$(R_2 \times 0.15 \, [\text{m}]) + (R_1 \times 0.85 \, [\text{m}]) = (100 \, [\text{N}] \times 0.35 \, [\text{m}]) + (50 \, [\text{N}] \times 1 \, [\text{m}])$$
$$85 [\text{Nm}] = 0.85 R_1 + 0.15 R_2 \qquad\qquad (2)$$

Now that we have two simultaneous equations that describe the forces acting on the beam we can solve R_1 and R_2.

$$65 \, [\text{Nm}] = 0.15 R_1 + 0.85 R_2 \qquad\qquad\qquad (1)$$

$$85 [\text{Nm}] = 0.85 R_1 + 0.15 R_2 \qquad\qquad\qquad (2)$$

We need an equation with only one unknown value in it to be able to calculate that value. If we subtracted one equation from the other the result would still contain two unknowns. If equation 1 is multiplied by 0.85/0.15,

$$\frac{0.85}{0.15} \times (65 \, [\text{Nm}] = 0.15 R_1 + 0.85 R_2)$$

$$368.333 \, [\text{Nm}] = 0.85 R_1 + 4.817 R_2 \qquad\qquad (3)$$

Call this equation 3. Notice now that both equation 3 and equation 2 contain a $0.15 R_1$ term. If we subtract equation 2 from equation 3:

$$(368.333 \, [\text{Nm}] - 85 \, [\text{Nm}]) = (0.85 R_1 - 0.85 R_1) + (4.817 R_2 - 0.15 R_2)$$

$$283.333 \, [\text{Nm}] = 4.667 R_2$$

$$R_2 = \frac{283.3 \, [\text{Nm}]}{4.67} = \mathbf{60.710 \, Nm}$$

Now we have a value for R_2, we can substitute this back into equation 1 to find a value for R_1.

$$65 \, [\text{Nm}] = 0.15 R_1 + (0.85 \times 60.710 \, [\text{Nm}]) \qquad\qquad (1)$$

$$R_1 = \frac{65 \, [\text{Nm}] - (0.85 \times 60.7 \, [\text{Nm}])}{0.15} = \mathbf{89.310 \, Nm}$$

We can check these answers by applying them to equation 2 and seeing if the equation is correct:

$$85\,[\text{Nm}] = 0.85R_1 + 0.15R_2 \qquad\qquad (2)$$
$$85\,[\text{Nm}] = (0.85 \times 89.310\,[\text{Nm}]) + (0.15 \times 60.710\,[\text{Nm}])$$
$$85\,[\text{Nm}] = 75.913\,[\text{Nm}] + 9.107\,[\text{Nm}]$$
$$85\,[\text{Nm}] = 85\,\text{Nm}$$

A **lever** is a simple machine that applies a force to a load by taking advantage of turning moments. It consists of a rigid bar that can pivot about a fulcrum. A force of effort is applied at one point and the lever then exerts a force against a load (Figure 2.42). The **mechanical advantage** is the ratio of the load to the effort. Consider a crow bar as shown.

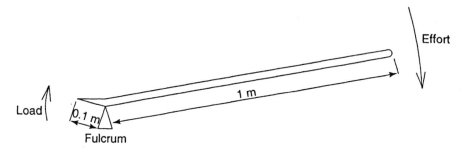

Figure 2.42

If an effort of 100 N is applied to the end of the handle, what force is applied to the load and what is the mechanical advantage?

$$\text{Clockwise moments} = \text{anti-clockwise moments}$$
$$100\,[\text{N}] \times 1\,[\text{m}] = \text{load force} \times 0.1\,[\text{m}]$$

$$\text{therefore load force} = \frac{100\,[\text{N}] \times 1\,[\text{m}]}{0.1\,[\text{m}]} = 1000\,\text{N}$$

$$\text{Mechanical advantage} = \frac{1000\,[\text{N}]}{100\,[\text{N}]} = 10$$

Notice that mechanical advantage has no units as the newtons cancel.

Another example of levers is in a pair of pliers or tin snips, as shown in Figure 2.43.

Here two levers operate about the same fulcrum. The forces of effort are applied by squeezing the handle causing one lever to act clockwise and the other anti-clockwise. The two forces act on the load in opposition to grip or snip the workpiece.

Distributed loads

So far all the loads on the beams have been concentrated, i.e. we have assumed that they are applied at an exact point on the beam and that the beam has a negligible mass. When the weight of the beam needs to be considered, or if a force does not act at a precise point, then we must assume a uniformly distributed load. This is a weight that has a certain value per unit length of the beam. We assume that this acts at its mid-point.

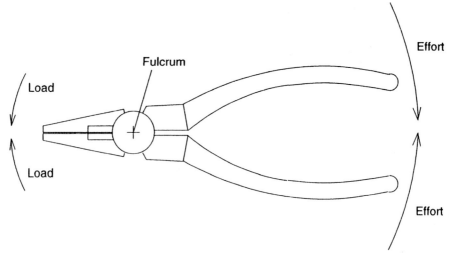

Figure 2.43 Pliers

☐ *Example 2.14*

The beam shown in Figure 2.44 pivots on a fulcrum at the far left end. A uniformly distributed mass of weight 25 N/m and length 1.5 m sits at the right end. An upward force of 20 N acts at the mid-point of the beam. Calculate the turning moment.

The weight of the uniformly distributed mass on the beam is calculated from the weight per length multiplied by the length:

$$25 \,[\text{N/m}] \times 1.5 \,[\text{m}] = 37.5 \,\text{N}$$

We must assume that the application point of the distributed load is at its mid-point, i.e. at 0.75 m from the right end where the arrow is. The force is then treated as any other.

$$\text{Clockwise turning moment} = 37.5 \,[\text{N}] \times 2.25 [\text{m}] = 84.375 \,\text{Nm}$$
$$\text{anti-clockwise turning moment} = -(20 \,[\text{N}] \times 1.5 \,[\text{m}]) = -30 \,\text{Nm}$$
$$\text{turning moment} = 84.375 \,[\text{Nm}] - 30 \,[\text{Nm}] = 54.375 \,\text{Nm}$$

As this is positive, then the resultant turning moment is **54.375 Nm acting clockwise**.

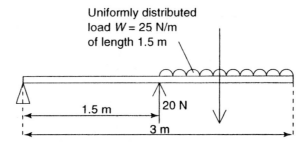

Figure 2.44 Crowbar

Centre of gravity

We already know that the weight of a body is defined as the force of attraction between it and the Earth. This definition does not say where the point of attraction is. Any body can be thought of as being made up of a large number of tiny particles. Each of these will have their own force of attraction to the Earth (see Figure 2.45).

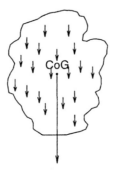

Figure 2.45

The resultant of all these forces is equal to the gravitational force on the body and it will act through a single point called the centre of gravity. A body acts as if all of its mass were concentrated at the centre of gravity. You can position the body anyway you want but the gravitational force will always act through this point.

Finding the position of the centre of gravity

Figure 2.46 shows a flat bar of negligible weight. On this bar there are various different masses, m_1, m_2 and m_3, positioned at distances of x_1, x_2 and x_3 from the left end of the bar.

We want to balance this bar on one support. For this to be possible, the support must be placed exactly under the centre of gravity. The resultant gravitational force

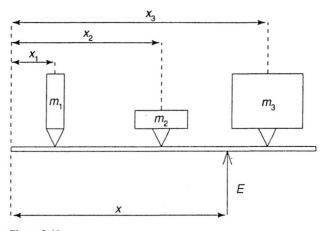

Figure 2.46

will act through the centre of gravity and its equilibrant will balance the bar. If we find the total sum of all the moments of force about one end of the bar then this will equal the moment of the equilibrant force about that end.

$$\text{The resultant gravitational force} = m_1g + m_2g + m_3g$$
$$= g(m_1 + m_2 + m_3)$$

If the equilibrant force is represented by E, and this acts at a distance of x from one end, the moment of this force will be Ex. Calculating moments about the left end:

$$(m_1g)x_1 + (m_2g)x_2 + (m_3g)x_3 = Ex$$
$$(m_1g)x_1 + (m_2g)x_2 + (m_3g)x_3 = g(m_1 + m_2 + m_3)x$$

Therefore, $x = \dfrac{(m_1g)x_1 + (m_2g)x_2 + (m_3g)x_3}{g(m_1 + m_2 + m_3)}$

The numerator of this equation is the summation of the moments of the gravitational forces, i.e. the weights, and the denominator is the summation of the weights. The word summation is represented by the Greek letter Σ, pronounced sigma. The formula above can be written in a general form as:

$$x = \frac{\Sigma \text{ moments of force}}{\Sigma \text{ force}}$$

You will notice that g appears as a multiple in the bottom and top of the equation and can be cancelled out. This leaves the expression:

$$x = \frac{\Sigma \text{ moments of mass}}{\Sigma \text{ mass}}$$

If all the masses considered are made of the same material or of materials that have the same density, since mass = density × volume, the density can be cancelled from every term, leaving volume in place of mass:

$$x = \frac{\Sigma \text{ moments of mass}}{\Sigma \text{ mass}}$$

$$\text{So, } x = \frac{\Sigma \text{ moments of (density} \times \text{volume)}}{\Sigma \text{ (density} \times \text{volume)}}$$

$$x = \frac{\Sigma \text{ moment of volume}}{\Sigma \text{ volume}}$$

Now imagine that all the masses not only have the same density, but are also made of plates or sections that all have the same thickness. Since volume = area × thickness, we can cancel thickness out of every term as well, leaving only area in place of mass:

$$x = \frac{\Sigma \text{ moment of volume}}{\Sigma \text{ volume}}$$

$$= \frac{\Sigma \text{ moment of (area} \times \text{thickness)}}{\Sigma \text{ (area} \times \text{thickness)}}$$

$$= \frac{\Sigma \text{ moment of area}}{\Sigma \text{ area}}$$

When only an area or a shape of a plate is considered then the answer x will show where the centre of gravity is. This would usually be called the 'centre of area' because area has no mass. The term 'centroid' is used when referring to the centre of area or centre of volume.

There are some common shapes with standard formulae for finding the centre of area and centre of gravity.

Centroid position of some common shapes

Rectangle
Point of intersection of the diagonals (Figure 2.47).
Triangle
Point of intersection of the medians. A median is a line from one of the vertices (corners) to the mid-point of the opposite sides (Figure 2.48).
Right-angled triangle
The centroid position is the same as for any triangle but in this case it is also 1/3 up and 1/3 along (Figure 2.49).
Circle
Point of intersection of any diameters (Figure 2.50).
Semi-circle
The distance from the diameter along the centre line, $X = (4R/3\pi)$, where R is the radius (Figure 2.51).

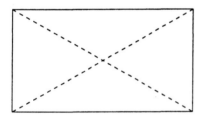

Figure 2.47

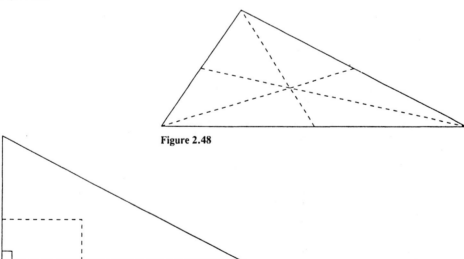

Figure 2.48

Figure 2.49

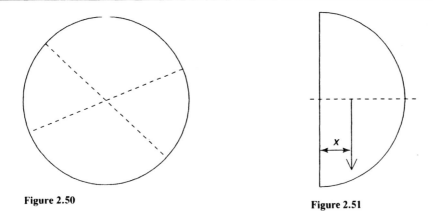

Figure 2.50   Figure 2.51

Centres of gravity of some common shapes

Cylinder or **rod**
Mid-point of the axis (Figure 2.52).
Solid cone or **pyramid**
The distance, $X = (h/4)$ (Figure 2.53).
Solid hemisphere
The distance from the diameter, $X = (3/8)R$, where R is the radius (Figure 2.54).

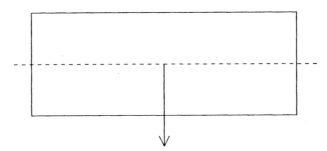

Figure 2.52

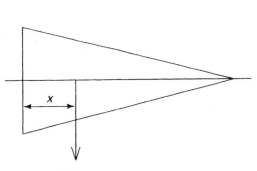

Figure 2.53

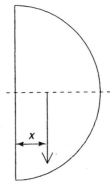

Figure 2.54

❏ *Example 2.15*

Find the position of the centroid of the plate shown in Figure 2.55.

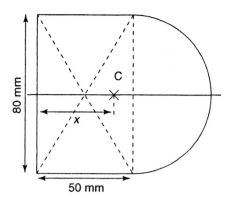

Figure 2.55

As this plate is symmetrical about the horizontal axis, straight away you know that the centroid C will lie on this line. As we are only considering the plate area and not the volume or mass, the formula to use is:

$$x = \frac{\Sigma \text{ moment of area}}{\Sigma \text{ area}}$$

Split the plate up into shapes, so that the areas can be easily calculated. In this case it is simple to split the plate up into a rectangle and a semi-circle. The next step is to choose one end of the plate as a reference and calculate the moments of the areas relative to this. We will choose the left hand end and consider the rectangle first.

$$\text{Area}_{(\text{rectangle})} = 80\,[\text{mm}] \times 50\,[\text{mm}] \qquad = 4000\ \text{mm}^2$$

$$\text{Moment of area}_{(\text{rectangle})} = 4000\,[\text{mm}^2] \times 25\,[\text{mm}] = 100\,000\ \text{mm}^3$$

Notice that the units of moment of area are mm³ which are the same as the units of volume, because moment of area is the product of three units of length.

Now look at the semi-circle. The area is fairly straightforward:

$$\text{Area}_{(\text{semi-circle})} = \frac{1}{2} \times \pi r^2 = \frac{\pi(40)^2}{2}\,[\text{mm}^2] = 2513.274\ \text{mm}^2$$

To find the moment of area about the left end, we need to multiply the area by the distance of the centroid of the semi-circle to the left end. From the standard case the centroid is ¾ of the radius from the straight edge and this needs to be added to the distance from the semi-circle edge to the left end of the plate:

$$\text{Moment of area}_{(\text{semi-circle})} = 2513.274\,[\text{mm}^2] \times \left(\left(\frac{4}{3\pi} \times 40 \right) + 50 \right) [\text{mm}]$$

$$= 168\,330.365\ \text{mm}^3$$

We can now apply the final formula:

$$x = \frac{\Sigma \text{ moment of area}}{\Sigma \text{ area}} = \frac{100\,000\,[\text{mm}^3] + 168\,330.365\,[\text{mm}^3]}{4\,000\,[\text{mm}^2] + 2513.274\,[\text{mm}^2]}$$

$$= 41.197\,\text{mm}$$

☐ *Example 2.16*

A vehicle has a wheelbase of 2.3 m and the distance from the front edge of the car to the front axle is 0.6 m. The load on the front axle is 3600 N and the load on the rear axle is 5500 N. Calculate the position of the centre of gravity. See Figure 2.56.

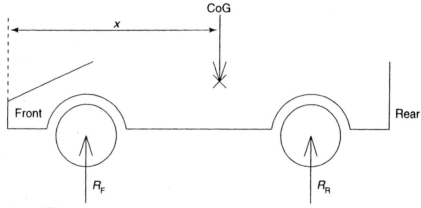

Figure 2.56

One method of solving this would be to take a balance of the moments:

clockwise moments = anti-clockwise moments

If the position of the centre of gravity from the front axle is a, and the total weight of the vehicle is 3600 N + 5500 N = 9100 N, then taking moments about the front axle:

$$a \times 9100\,[\text{N}] = 5500\,[\text{N}] \times 2.3\,[\text{m}]$$

$$\text{Therefore } a = \frac{5500\,[\text{N}] \times 2.3\,[\text{m}]}{9100\,[\text{N}]} = \mathbf{1.390\,m}$$

Alternatively, if the distance of the centre of gravity to the front edge of the car is b then using the front edge of the car as an axis:

$$b = \frac{\Sigma \text{ moments of force}}{\Sigma \text{ force}}$$

$$= \frac{(3600\,[\text{N}] \times 0.6\,[\text{m}]) + (5500\,[\text{N}] \times 2.9\,[\text{m}])}{(3600\,[\text{N}] + 5500\,[\text{N}])}$$

$$= 1.990\,\text{m}$$

The distance of the centre of gravity from the front axle:

$1.990\,[\text{m}] - 0.6\,[\text{m}] = 1.390\,\text{m}$, which agrees with the first method.

❑ *Example 2.17*

Find the position of the centre of gravity of the shaft shown in Figure 2.57.

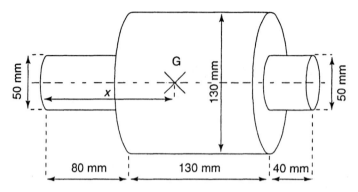

Figure 2.57

Split the shaft up into easy to calculate shapes, i.e. three cylinders. Pick one end, e.g. the left, and calculate the centre of gravity relative to this. We could use the formula:

$$x = \frac{\Sigma \text{ moments of mass}}{\Sigma \text{ mass}}$$

but if we assume that the shaft is made of all the same material then the density will be the same for all the shaft and so we can use the formula:

$$-x = \frac{\Sigma \text{ moment of volume}}{\Sigma \text{ volume}}$$

We can use the formula $\quad V = \dfrac{\pi d^2}{4} \times l$ or $\pi r^2 \times l$ for for all three cylinders.

$$\text{Volume}_{(\text{cylinder 1})} = \frac{\pi(50)^2}{4} \times 80 = 157.080 \times 10^3 \text{ mm}^3$$

Now we can calculate the moment of volume about the left end:

$$\text{Moment of volume}_{(\text{cylinder 1})} = 157.080 \times 10^3 \, [\text{mm}^3] \times 40 \, [\text{mm}]$$

$$= 6283.185 \times 10^3 \text{ mm}^4$$

$$\text{Volume}_{(\text{cylinder 2})} = \frac{\pi(130)^2}{4} \times 130 = 1725.520 \times 10^3 \, [\text{mm}^3]$$

$$\text{Moment of volume}_{(\text{cylinder 2})} = 1725.520 \times 10^3 \, [\text{mm}^3] \times \left(80 + \frac{130}{2}\right) [\text{mm}]$$

$$= 250\,200.366 \times 10^3 \text{ mm}^4$$

$$\text{Volume}_{(\text{cylinder 3})} = \frac{\pi(50)^2}{4} \times 40 = 78.540 \times 10^3 \text{ mm}^3$$

$$\text{Moment of volume}_{(\text{cylinder 3})} = 78.540 \times 10^3 \, [\text{mm}^3] \times \left(80 + 130 + \frac{40}{2}\right) [\text{mm}]$$

$$= 18\,064.158 \times 10^3 \, \text{mm}^4$$

$$= 18\,064.158 \times 10^3 \, \text{mm}^4$$

$$x = \frac{\Sigma \text{ moment of volume}}{\Sigma \text{ volume}}$$

$$x = \frac{(18\,064.158 + 250\,200.366 + 6283.185) \times 10^3 \, [\text{mm}^4]}{(157.080 + 1725.520 + 78.540) \times 10^3 \, [\text{mm}^3]}$$

$$= \frac{274\,547.7 \times 10^3 \, [\text{mm}^4]}{1961.14 \times 10^3 \, [\text{mm}^3]}$$

$$= \mathbf{139.994 \, mm}$$

So the position of the centre of gravity is 140 mm from the left end.

2.4 Frameworks

Bow's notation

Bow's notation is a method of labelling forces in space diagrams and force diagrams. In a space diagram the spaces are labelled with capital letters, as shown in Figure 2.58.

Each force can then be referred to by the letters of the two spaces either side of the force (Figures 2.59 and 2.60).

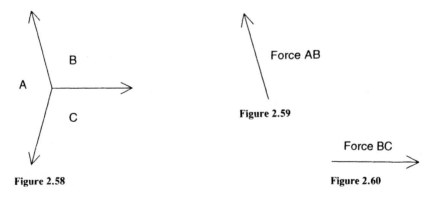

Figure 2.58

Figure 2.59

Force AB

Force BC

Figure 2.60

The force is represented on the force diagram by a vector labelled by the corresponding lower case letters on the ends of the vectors, as shown in Figure 2.61.

This method simplifies problems when dealing with a whole system of forces. The lettering on the space diagram can be clockwise or anti-clockwise but it is advisable to stick to one direction.

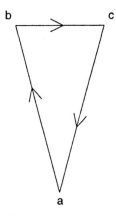

Figure 2.61

Framed structures

A framed structure is a structure built of straight bars joined at the ends. The subject of frameworks is concerned with applying the principles of equilibrium of forces, and stress and strain, to structures to measure their internal forces. Then the material dimensions required can be calculated. It is assumed that the ends are joined by frictionless pins, so that each bar is in direct tension or compression. In practice the bars may be actually be riveted or welded, but this assumption simplifies the theory.

A bar in compression is called a strut. The compressive force is resisted by an internal force that 'pushes' towards the ends.

A bar in tension is called a tie and the internal resisting force 'pulls' the ends. It is assumed that the force acting in any bar acts in the same physical direction as that bar.

Framed structures are sometimes called light frameworks, as the weight of each member is a much lower force than the force that the member carries. Frameworks are generally made up of triangular shaped frames: take a look at bridges, roofs and cranes. In order that an engineer can design the frame, the forces that act on each frame member must be known.

So far the force systems we have looked at have mainly considered forces that meet at one particular node (the crane head of the jib crane and the gudgeon pin of the reciprocating engine mechanism). In framed structures, the forces that meet at every node are considered. Some of these forces are internal and others are external such as loads and reactions. If the forces that meet at any node of a framework are considered, we have a force system in equilibrium and a closed vector diagram can be drawn for that node. We will only consider systems of co-planar forces in this chapter, i.e. frameworks in one plane

To recap, if a body is under the action of a system of co-planar forces and is in equilibrium then:

- the algebraic sum of the components of the forces in any two directions must be zero, implying that there is no linear acceleration
- the algebraic sum of the moments of all the forces about any node must be zero, implying that there is no angular acceleration.

To make provision for expansion due to temperature changes, one end of the framework is usually supported by a hinge and the other by rollers. The reaction of a roller support is always assumed to be at right angles to the surface in contact, as shown in Figure 2.62.

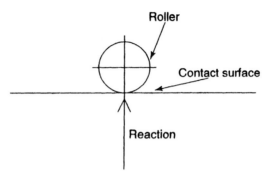

Figure 2.62

The reaction of the hinge can be in any direction but if the external loads are vertical and the roller reactions are vertical then the hinge reactions will also be vertical. Framed structure problems can be solved by the following vector diagram methods if the framework is only loaded at the joints.

Space diagrams are based on the framework layout.

❏ *Example 2.18*

Consider a roof truss. A load of W is supported at the apex. The two reaction supports are

$$\frac{W}{2}$$

Draw the vector diagram for any value of W.

Using Bow's notation we label all the spaces with capital letters, A,B,C and D for example (Figure 2.63). We will consider the letters in a clockwise direction. If each node is considered in turn we will have three separate vector diagrams.

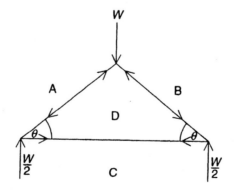

Figure 2.63

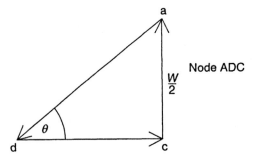

Figure 2.64 Node ADC

Consider node ADC (Figure 2.64). The direction in which the forces act can be logically determined. The reaction CA must act vertically upwards. The vertical component of AD must act downwards. AD has a horizontal component acting to the left. Therefore to keep the system in equilibrium the force of DC must act horizontally to the right. Similarly for node ABD (Figure 2.65) and node BCD (Figure 2.66).

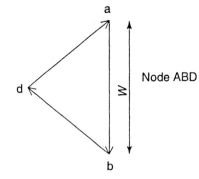

Figure 2.65 Node ABD

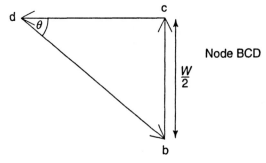

Figure 2.66 Node BCD

In any one member, the arrows on the space diagram are in opposite directions, as each member is in either direct compression or tension. Because of this, if we want to combine the above three diagrams then the arrows are omitted as they would be in opposition to each other (Figure 2.67).

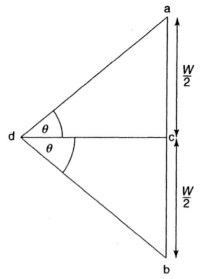

Figure 2.67 Combined diagram

The combined diagram is really diagrams BCD and ADC superimposed on ABD. After a little practice you will not usually need to draw the vector diagram for each node: you will be able to go straight into the combined vector diagram for the whole structure.

☐ *Example 2.19*

Draw the complete vector diagram for the framework shown in Figure 2.68.

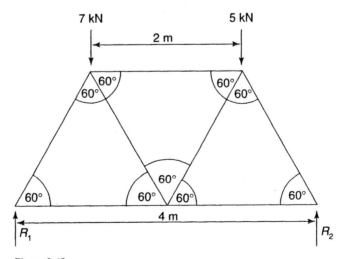

Figure 2.68

To find R_1 and R_2:

About R_1

$$\text{clockwise moments} = \text{anti-clockwise moments}$$

$$(7\,[\text{kN}] \times 1\,[\text{m}]) + (5\,[\text{kN}] \times 3\,[\text{m}]) = R_2 \times 4\,[\text{m}]$$

$$22 = 4\,R_2$$

$$\text{Therefore } R_2 = \frac{22}{4}\left[\frac{\text{kNm}}{\text{m}}\right] = 5.5\,\text{kN}$$

About R_2

$$\text{clockwise moments} = \text{anti-clockwise moments}$$

$$R_1 \times 4\,[\text{m}] = (7\,[\text{kN}] \times 3\,[\text{m}]) + (5\,[\text{kN}] \times 1\,[\text{m}])$$

$$26 = 4\,R_1$$

$$\text{Therefore } R_1 = \frac{26}{4}\left[\frac{\text{kNm}}{\text{m}}\right] = 6.5\,\text{kN}$$

Select a suitable scale and draw the space diagram. Show the direction of the external forces. Use Bow's notation to label the space between the forces.

Look at Figure 2.69. All the external forces act vertically. Draw these first. The force ab is 7 kN and acts downwards. The force bc is 5 kN and acts downwards. The reaction force CD, 5.5 kN, acts upwards from c to d. So far, this is just a straight line but the points a, d, b and c are known. The adjoining spaces to E are A and D. The points a and d are already known on the force diagram. Draw a line parallel to AE through a. Draw a line parallel to ED through point d. The intersection gives the point e. The line ed represents the force ED and the line ae represents the force AE.

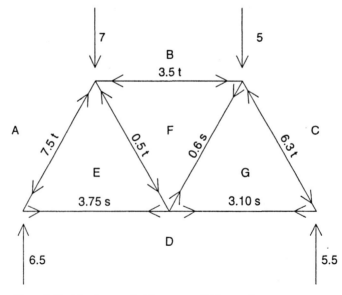

Figure 2.69 All values are in kilonewtons (kN): t = tie; s = strut

Through e draw ef parallel to EF. Through point b draw bf parallel to BF. The intersection gives point f.

Through c draw cg parallel to CG. Through d draw line dg parallel to DG. The intersection gives point g. The line gf can then be drawn parallel to GF (Figure 2.70).

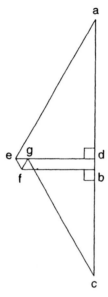

Figure 2.70

Measure the lengths of the lines on the force diagram to establish the various forces. The direction of the forces can be obtained by considering each node in turn and working clockwise around the node. For example, for node ABFE, see Figure 2.71.

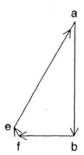

Figure 2.71

Each node forms a closed vector diagram, but the arrow direction can be determined by following the appropriate force around the combined diagram. You will then have the magnitude of the forces in all members of the framework and will know whether they are in tension or compression. Then you can indicate the magnitude and nature of the forces as shown on the space diagram.

Method of sections

The method of sections is used when we just want to find the forces in a particular member rather than all the members of a framed structure. The framework is imagined to be divided into two parts, the dividing line cutting the member under investigation. As frameworks are generally made up of triangular-shaped frames, then the dividing line also cuts two other members. The part of the framed structure to one side of the dividing line will then only remain in equilibrium if three forces are added that are normally provided by the rest of the structure. The value of the unknown force of the member can then be found by taking moments about the intersection of the other two unknown forces. The other two unknown forces have a zero moment about a point as they pass through it, leaving one unknown in a moment equation.

☐ *Example 2.20*

Consider the last example. Find the forces in members ED and BF using methods of sections and check your answers with the previous force diagram.

To find the force in ED imagine the structure divided as in Figure 2.72.

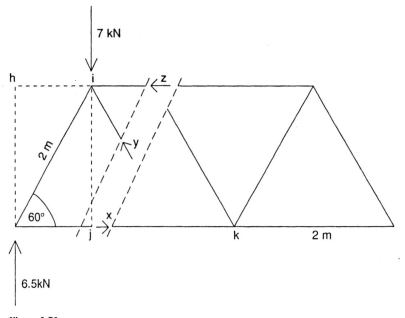

Figure 2.72

The left part of the structure will remain in equilibrium if three forces are added, x, y and z, previously provided by the members of the structure that the line divides. The left part of the structure is therefore under the influence of five forces: two known (7 kN and 6.5 kN) and three unknown (x, y and z) forces. The force x is the force in ED that we need to find. The other two unknown forces intersect at point i on the diagram. The value of x can be found by taking moments about the intersection.

clockwise moments = anti-clockwise moments

$$6.5\,[\text{kN}] \times \text{hi} = (x \times \text{ij})$$

To find length ij,

$$\sin 60° = \frac{ij}{2}$$

Therefore ij $= \sin 60° \times 2 = 1.732$ m

$$\text{length hi} = \frac{2\,[m]}{2} = 1 \text{ m}$$

$$6.5\,[kN] \times 1\,[m] = (x \times 1.732\,[m])$$

$$\text{Therefore } x = \frac{6.5}{1.732}\left[\frac{kNm}{m}\right] = \mathbf{3.75\ kN}$$

To find the force in BF, z, take moments about the intersection of forces x and y.

clockwise moments = anti-clockwise moments

$$6.5\,[kN] \times 2\,[m] = (7\,[kN] \times 1\,[m]) + (z \times 1.732\,[m])$$

$$13\,[kNm] = 7\,[kNm] + 1.732z$$

$$\text{Therefore } z = \frac{(13-7)}{1.732}\left[\frac{kNm}{m}\right] = \mathbf{3.5\ kN}$$

☐ *Example 2.21*

Find the force in member CD in Figure 2.73 using the method of sections.

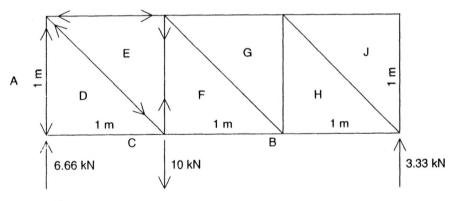

Figure 2.73

Take moments about the intersection of x and y to find z (Figure 2.74).

clockwise moments = anti-clockwise moments

$$(6.66\,[kN] \times 0) = (z \times 1\,[m])$$

$$\text{Therefore } z = \frac{(6.66 \times 0)}{1}\left[\frac{kNm}{m}\right] = \mathbf{0}$$

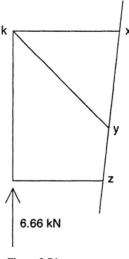

Figure 2.74

The force in member DC is therefore zero. A member of a framed structure with a zero force acting in it is called a **redundant** member. A redundant member provides *no* support to the loaded structure.

Problems 2

1. A car has a mass of 1600 kg. What is the weight on each tyre, assuming that these are equal?
2. The dimensions of a fuel tank are 0.5 m by 0.3 m by 0.35 m. What mass of fuel does this hold of relative density 0.75?
3. If the mass of an engine cylinder head is 50 lb, what is its weight? If the density of the material is 7900 kg/m³, then what volume material is required to make the cylinder head? Neglect any extra material that would be machined off during manufacture.
4. A truck stands at rest. A person with a mass of 75 kg gets into the truck, goes to fill up with 10 gallons of fuel with a relative density of 0.9 and then loads 0.25 tonne of scrap steel on the back. What is the increase in weight of the vehicle, neglecting any fuel used in the journey?
5. The weight of a milk float reduces by 1716.75 N during an hour of its morning round. If all the bottles are 0.5 litres in size then how many bottles are delivered during the hour? Assume the density of the milk to be the same as water and that the bottles are made of plastic of negligible weight.
6. A car with a mass of 2822 lb has a flat battery. A person with a mass of 75 kg gets into the car to steer it and two people apply a force of 400 N to the back of the car to bump start it. What acceleration does the car have, assuming friction to be negligible?
7. A car stands at rest on a cold winter morning. A man weighing 90 kg jumps into the car and drives off. A few minutes later the radiator bursts and he realises that he's forgotten to put antifreeze into the coolant system. 4.2 litres of water leak

out of the radiator. What is the change in weight of the car from before the driver got in?

8. The bore of a four cylinder petrol engine is 100 mm and the stroke is 86 mm. What is the capacity of the engine?

9. A piston has a mass of 0.75 kg. The force on the gudgeon pin at an instant in time is 650 N. Calculate the acceleration of the piston.

10. A tank holds 0.07 m³ of oil of relative density 0.8. What volume of petrol would this tank hold of relative density 0.75?

11. A four wheel vehicle has a mass of 1500 kg. Find the force that acts between the road and tyres if the force on each tyre is equal.

12. A 27 kN force acts at an angle of 27° to the horizontal. Find the horizontal and vertical components.

13. A car is towed by a rope. The force exerted by the rope is 1000 kN but is at an angle of 10° to the forward direction of the car. What is the effective force pulling the car forward?

14. A motor vehicle has a mass of 1350 kg and rests on a slope of 10°. Calculate the normal reaction between the road and the tyres. What is the force parallel to the road surface to keep the car stationary?

15. An engine of mass 200 kg is suspended by a chain from an overhead crane. Someone leans on the engine and exerts a horizontal force. The crane takes up a new position 30° to the vertical. What are the forces on the chain and the horizontal force?

16. A car is pushed by a force of 1000 N. The mass of the car is 900 kg. Find the resultant force of the weight and the 'push'.

17. A system of co-planar forces acts on a body. The equilibrant force is also applied to the body. Why does the body not move?

18. A wall crane has a jib AB hinged at the wall at A and is supported by a tie CB. The angle between the jib and the wall is 140° and the angle between the tie and the wall is 105°. A mass of 1020 kg is supported from the jib at B. Determine what forces act in the tie CB and the jib AB. Assume $g = 9.81$ m/s².

   Hint: Start by drawing the space diagram. Draw a vertical line to represent the wall and label it AC. Next draw a line representing the jib upwards to the right at an angle of 140° to AC and label it AB. Then draw a line representing the tie above this at an angle of 105° to AC. In this case, the scale of the space diagram is not important because you do not need to measure any angles or positions on the space diagram to use on the force diagram. Calculate the gravitational force on the mass using *mg* and draw a line vertically down from B to represent this. Put information that you already know on the diagram in an appropriate place, such as the angles and the weight of the mass. You now have a clear picture of the situation and can draw the force diagram. Select a scale that makes the weight *mg* about half the length of your paper. Draw the force *mg* vertically on the lower left part of your paper. Draw a line from the bottom of this at an angle of 140°. Take it right across your paper because you do not yet know how long it needs to be. Now you can draw the final line in from the top of the gravitational force line at an angle of 105°. Take this line right across the paper. You should now have the force triangle and can measure the force lines of the jib and tie. The gravitational force on the mass, *mg*, equals 10.006 kN.

19. The piston of a reciprocating engine exerts a force of 10 kN at the top of the piston when the crank is 35° past the top dead centre (TDC). If the stroke of the piston is 100 mm and the length of the connecting rod is 165 mm, find the guide force and the force in the connecting rod.
    Hint: Start by drawing a space diagram to scale. You know the length of the connecting rod. The length of the crank can be found from the stroke length. If the crank is drawn at an angle of 35°, then the angle between the connecting rod and the vertical can be measured. This angle can then be used to draw the force diagram.

20. A petrol engine has a connecting rod 200 mm long and a stroke of 120 mm. When the crank is 40° past TDC the force on the piston is 5 kN. Calculate the force on the connecting rod, the guide force, and the angle between the connecting rod and the line of the piston stroke.

21. A torque of 33.75 Nm is applied to a socket and ratchet. The length of the handle is 225 mm. If the handle is changed for a longer one of length 350 mm and the same force is applied to the end of the handle, calculate the new torque produced.

22. A steel bar is used as a lever to apply a load force. The bar is 1.5 m long. The pivot is positioned 0.3 m from the load end of the bar. What effort is required to apply a load force of 700 N?

23. A bar is 500 mm long and carries loads of 20 N, 55 N and 80 N at distances of 100 mm, 250 mm and 450 mm respectively from the left end. Neglect the weight of the bar and calculate the position of a single support placed so that the bar would balance.

24. In an overhead valve engine, the push rod exerts a force of 300 N on the rocker. The distance from the fulcrum to the push rod centre is 70 mm. The distance from the fulcrum to the valve centre is 65 mm. Calculate the force on the valve at this instant.

25. A vehicle has a wheel base of 3.9 m. The load on the front axle is 8000 N and the load on the rear axle is 9500 N. Find the centre of gravity.

26. A vehicle has a total weight of 12 900 N and the centre of gravity is 1.020 m from the rear axle. The length of the wheel base is 2.8 m. If the front wheels are to be lifted clear of the ground by a hydraulic jack placed under the front axle, what force must the jack exert?

27. A commercial vehicle has a wheel base of 3.5 m. The load on the front axle is 9000 N and the load on the rear axle is 12 000 N. A load of 2000 N is placed in the vehicle at a distance of 2.9 m from the front axle. By how far does the centre of gravity move when the extra load is added?

28. A beam is 1.25 m long. However, it is not of constant section. 500 mm of the beam from the left end the way it is positioned acts as a uniformly distributed load of 120 N/m. The rest of the beam acts as a uniformly distributed load of 70 N/m. Calculate the position of the centre of gravity.

29. A steel shaft consists of three sections, all concentric, i.e. all have the same axis, and made of the same steel throughout. As it is positioned the left section is 90 mm long and has a diameter of 65 mm; the middle section is 65 mm long and has a diameter of 85 mm; and the right end section is 80 mm long and has a diameter of 60 mm. Find the position of the centre of gravity.

30. A plate measures 1.1 m long by 0.75 m wide. A hole of 0.1 m diameter is drilled on the lengthways centre line 0.3 m from the left end. Calculate how far the centre of area moves when the hole is drilled.

Hint: As the plate is symmetrical about the lengthways centre line, then you only need to consider the position of the centre of area from one edge, x, i.e. in one direction. Consider that the hole drilled has a negative area and a negative moment of area and subtract these values from those of the rest of the plate.

3 Distortion of materials

3.1 Stress and strain, and strength of materials

Stress and strain

Stress and **strain** are common everyday terms but in engineering they have specific meanings. Stress and strain occur in a material when a load is applied and they are closely related. A load is any force acting on a body, in tension or compression. Every component of a vehicle and engine has many loads acting on it. Besides the component's own weight, there are also loads such as centrifugal forces, frictional forces and drive forces. The definition of stress is 'the internal resistance set up in a material when its shape is changed by the application of an external force'. This is not easy to understand on first reading but you will get a clearer picture of it as you work through this section.

There are three different types of stress to look at: **compressive stress, tensile stress** and **shear stress**. For the moment we'll consider compression and tension.

When a load tends to squash or shorten material, it is said to be in compression (Figure 3.1). When a load tries to stretch a material it is said to be in tension (Figure 3.2). Both these loads produce a type of stress called **direct stress**. It is measured as the force applied per unit of cross-sectional area. The cross-sectional area must always be taken perpendicular to the line of action of the force. The quantity symbol for stress is σ. Its units are derived from force per unit area and so are N/m^2, the same as the units of pressure.

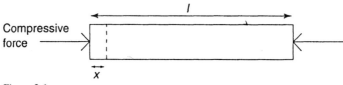

Figure 3.1

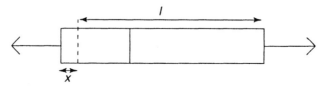

Figure 3.2

$$\text{Direct tensile stress, } \sigma_t = \frac{\text{tensile force}}{\text{cross-sectional area}} = \frac{F}{A} \left[\frac{N}{m^2} \right]$$

$$\text{Direct compressive stress, } \sigma_c = \frac{\text{compressive force}}{\text{cross-sectional area}} = \frac{F}{A} \left[\frac{N}{m^2} \right]$$

When a material is under direct stress like this, a change in shape takes place. A compressive force will tend to contract a material in the direction of the force and make it expand perpendicular to the direction of the force, i.e. the force will make the material shorter and fatter. A tensile stress will tend to expand the material in the direction of the force and contract it perpendicular to the force, i.e. the stress will make the material longer and thinner. It is this change of shape that is called strain.

Linear strain is a measure of the deformation that takes place when a material is loaded in compression or tension. It is the change of length per unit length measured in the direction that the force acts. The quantity symbol of linear strain is ε and has no units, as length is divided by length.

$$\text{Linear strain, } \varepsilon = \frac{\text{change of length}}{\text{original length}} = \frac{x}{l}$$

where x is the change in length and l is the original length. The units of length used for x and l do not matter, so long as they are the same.

When a material is under stress, a corresponding strain must occur. If the load is removed and the material returns to its original shape (the strain is totally removed), then the material is said to be *perfectly elastic*. This property of a material is known as **elasticity**. All of the calculations we will do are based on the assumption that a material remains perfectly elastic and obeys **Hooke's law**.

Hooke's law states that: 'stress is directly proportional to the strain produced, provided that the limit of proportionality is not exceeded'. We'll come back to the limit of proportionality later. For the moment remember that you can assume that stress is proportional to strain so long as the force is not too great.

$$\text{As stress} \propto \text{strain, then stress} = \text{strain} \times \text{constant}$$

$$\text{and constant} = \frac{\text{stress}}{\text{strain}}$$

This constant is the same value for any given material and the relationship is useful for calculations. For direct stresses, it is called **Young's modulus of elasticity** and usually has the quantity symbol E, so:

$$\frac{\text{direct stress}}{\text{direct strain}} = \text{Young's modulus of elasticity}$$

$$\frac{\sigma}{\varepsilon} = E$$

Young's modulus of elasticity is a measure of the resistance a material has to tensile and compressive forces. The higher the value of this modulus E the less a material will extend when a given load is applied. Table 3.1 shows some typical values of E for some common engineering materials.

Table 3.1 Typical values of Young's modulus of elasticity

Material	$E[GN/m^2]$
Mild steel	210
Phosphor bronze	90
Grey cast iron	117
Brass	100

❑ *Example 3.1*

A mild steel engine bolt when correctly tightened provides a tensile load of 3 kN. The bolt is 300 mm long and has a diameter of 15 mm. Calculate the stress on the material. How much will the bolt extend? For mild steel $E = 200 \times 10^9$ N/m^2 = 200 GN/m^2.

To calculate the stress we need to know the force and the cross-sectional area of the bolt.

$$\text{Area, } A = \frac{\pi d^2}{4} = \frac{\pi (15 \times 10^{-3})^2}{4} \ [\text{m}^2]$$

$$= 176.715 \times 10^{-6} \ \text{m}^2$$

$$\sigma = \frac{F}{A} = \frac{3 \times 10^3}{176.715 \times 10^{-6}} \left[\frac{\text{M}}{\text{m}^2}\right]$$

$$= 16.976 \times 10^6 \left[\frac{\text{N}}{\text{m}^2}\right]$$

$$= 16.976 \ \text{MN/m}^2$$

As $\varepsilon = x/l$, then $x = \varepsilon \times l$. We know l, the original length, so if we can calculate ε we can find out x, the extension of the bolt.

As $E = \sigma/\varepsilon$, then $\varepsilon = \sigma/E$. We can use this formula for ε in the formula for x and then be able to calculate the extension.

$$x = \varepsilon \times l = \frac{\sigma}{E} \times l$$

$$= \frac{16.976 \times 10^6}{200 \times 10^9 \times 300 \times 10^3} \left[\frac{\text{N}}{\text{m}^2}\right]\left[\frac{\text{m}^2}{\text{N}}\right][\text{m}]$$

We can cancel out the 10^x multiples next:

$$\frac{10^6 \times 10^{-3}}{10^9} = \frac{10^3}{10^9} = \frac{1}{10^6} = 10^{-6}$$

$$\text{Thus } \frac{16.976 \times 10^{-6}}{200} \times 300 \ [\text{m}] = 25.464 \times 10^{-6} \ \text{m}$$

$$= \textbf{25.564 } \boldsymbol{\mu}\textbf{m or 0.0255 mm}$$

Tensile tests

Tensile tests are carried out to find out how a material behaves under different tensile loads. A standard test is used so that comparisons can be made. A test piece of a material of dimensions fixed by a British Standard Specification is gradually pulled by a testing machine until it breaks. At frequent intervals during the test, the length and corresponding load are recorded. A graph can then be drawn, called a load–extension diagram, showing the characteristics of the material. Each metal has a characteristic graph. The shape of this will depend also on its composition. A typical diagram for mild steel looks like that shown in Figure 3.3.

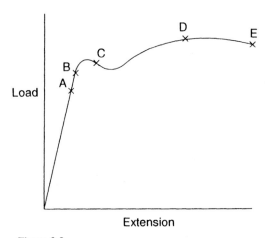

Figure 3.3

The important points on this diagram are as follows:

A is the limit of proportionality that was mentioned earlier. Up to this point Hooke's law applies. The line to this point is straight and so the basic formula can be applied.
B is the elastic limit. If the material is stretched any more than this, it will not return to its original dimensions.
C is the yield point, where a sudden increase in length occurs without any increase in load.
D is the ultimate load. This is the maximum load recorded during the test.
E is the breaking point. Notice that the load required for this is less than the ultimate load because the test piece starts to narrow at the place where it will break.

Not all materials have the linear relationship up to the limit of proportionality and for these materials Hooke's law does not apply. Fortunately for us, steels do exhibit this linearity and the formulae above can be used.
There are various stresses that we need to define for the test:

$$\text{ultimate tensile stress} = \frac{\text{maximum load}}{\text{original CSA}}$$

where CSA is the cross-sectional area. (Ultimate tensile stress is sometimes called tensile strength.)

$$\text{nominal breaking stress} = \frac{\text{load at fracture}}{\text{original CSA}}$$

$$\text{real breaking stress} = \frac{\text{load at fracture}}{\text{CSA at fracture}}$$

The two different values for breaking stress are used because the cross-sectional area at fracture is less than the original so the real breaking stress is greater than the nominal.

The **factor of safety** is an important consideration for engineering design. It is the ratio of the ultimate stress of a material to the safe allowable working stress to which it may be subjected.

$$\text{Factor of safety} = \frac{\text{ultimate stress}}{\text{safe working stress}}$$

The safe working stress is the maximum permissible stress for any component. This is decided by considering a number of factors, as follows:

1   The type and condition of the material; e.g. are there likely to be any weaknesses or flaws?
2   The effect of wear and corrosion.
3   Whether the load is a dead weight, or one gradually applied or one suddenly applied.
4   The effects of failure.
5   How well the component is manufactured.

The factor of safety may vary from about 3 for simple static loads to about 20 or more for complex shock loads, where the effects of failure may be disastrous.

Two other definitions that are useful factors are percentage elongation and percentage reduction in area:

$$\text{percentage elongation} = \frac{\text{length at fracture} - \text{original length}}{\text{original length}} \times 100\%$$

$$\text{percentage reduction in area} = \frac{\text{original CSA} - \text{CSA at fracture}}{\text{original CSA}} \times 100\%$$

❏ *Example 3.2*

A tie bar 5 m long has a CSA of 415 mm^2 and is designed to support an axial load of 50 kN. If the ultimate tensile strength for the material is 500 MN/m^2 and $E = 200$ GN/m^2, calculate

(a) the stress in the bar
(b) the strain
(c) the factor of safety at this load.

$$\sigma = \frac{F}{A} = \frac{50 \times 10^3}{415} \left[\frac{N}{mm^2} \right] \times 10^6 \left[\frac{mm^2}{m^2} \right]$$

$$= 120.482 \times 10^6 \ [N/m^2] = \mathbf{120.482 \ MN/m^2}$$

As $E = \dfrac{\sigma}{\varepsilon}$, then $\varepsilon = \dfrac{\sigma}{E}$

$$\varepsilon = \frac{120.482 \times 10^6}{200 \times 10^9} \left[\frac{\cancel{N}}{\cancel{m^2}}\right]\left[\frac{\cancel{m^2}}{\cancel{N}}\right] = 0.602 \times 10^{-3}$$

$$\text{factor of safety} = \frac{\text{ultimate stress}}{\text{safe working stress}}$$

$$= \frac{500}{120.482} \left[\frac{\cancel{MN}}{\cancel{m^2}}\right]\left[\frac{\cancel{m^2}}{\cancel{MN}}\right] = 4.1$$

Resilience

The work done by a force can be calculated from the product of the average force applied and the distance the force moves through. As we have seen, if a direct force is applied to a material then the material deforms. Work is done to strain the material and the energy is then stored in the material, provided that the elastic limit is not exceeded. The energy stored is called **resilience**. This resilience can be estimated by calculating the work done in straining the material. Consider a load gradually applied to a material from 0 to F N. The average load is therefore

$$\frac{0 + F}{2} = \frac{F}{2}\ \text{N}$$

If the deformation is measured by x m, (compression or tension), then the work done to strain the material is:

$$\text{resilience} = \text{force} \times \text{distance}$$

$$= \frac{F}{2}\,[\text{N}] \times x\,[\text{m}] = \frac{1}{2}Fx\ [\text{Nm}]$$

$$\text{Maximum stress, } \sigma = \frac{F}{A}$$

$$\text{then } F = \sigma \times A$$

$$\varepsilon = \frac{x}{l}$$

$$\text{then } x = \varepsilon \times l$$

$$\text{Resilience} = \frac{1}{2}Fx\ [\text{Nm}]$$

$$= \frac{1}{2} \times \sigma \times A \times \varepsilon \times l$$

$$\text{As } E = \frac{\sigma}{\varepsilon}$$

$$\text{then } \varepsilon = \frac{\sigma}{E}$$

therefore resilience $= \dfrac{1}{2} \times \sigma \times A \times \dfrac{\sigma}{E} \times l$

As volume, $V = A \times l$

then resilience $= \dfrac{\sigma^2 V}{2E}$

Check units: $\left[\left(\dfrac{N}{m^2}\right)^{\cancel{2}}\right][m^{\cancel{3}}]\left[\dfrac{m^2}{\cancel{N}}\right] = [Nm]$

❑ *Example 3.3*

Calculate the resilience in Example 3.2.

resilience $= \dfrac{\sigma^2 V}{2E}$

$= \dfrac{(120.482 \times 10^6) \times 415 \times 10^{-6} \times 5}{2 \times 200 \times 10^9}\left[\left(\dfrac{N}{m^2}\right)^{\cancel{2}}[m^{\cancel{3}}]\left[\dfrac{m^2}{\cancel{N}}\right]\right]$

$= 75.301 \text{ Nm} = 75.301 \text{ J}$

Compound bars

A compound bar is any structural component made of two or more materials arranged in parallel, rigidly connected together at the ends with an axial load. Think of a rod and a sleeve of the same length but of different materials (Figure 3.4). This complicates things a bit so we need to adapt the stress and strain formulae.

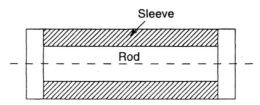

Figure 3.4 Compound bar

Consider a comprehensive load applied to this compound bar. The load is shared between the rod and the sleeve. The sum of the loads carried by the rod and sleeve must equal the total applied load.

As $\sigma = \dfrac{F}{A}$ then $F = \sigma A$. For the compound bar then:

$$F_{total} = F_{tube} + F_{sleeve}$$

therefore $F_{total} = \sigma_{tube} A_{tube} + \sigma_{sleeve} A_{sleeve}$

Since the rod and the sleeve are initially the same length and they will shorten by the same amount under the load then their strains are equal:

$$\frac{x}{l} = \varepsilon_{sleeve} = \varepsilon_{rod}$$

As $E = \sigma/\varepsilon$ while the loading is within the limit of proportionality then $\varepsilon = \sigma/E$, and so:

$$\frac{\sigma_{sleeve}}{E_{sleeve}} = \frac{\sigma_{rod}}{E_{rod}}$$

Therefore, $\dfrac{\sigma_{sleeve}}{\sigma_{rod}} = \dfrac{E_{sleeve}}{E_{rod}}$

The stresses induced are in the same ratio as the values of the Young's modulus of elasticity. The above formulae will enable the different stresses in compound bars to be calculated. They will also apply to tensile loads.

☐ *Example 3.4*

A steel bar and brass bar each have a length of 500 mm and a rectangular cross-section 13 mm by 12.5 mm. They are bonded together to form a composite bar 500 mm long with a cross-section 13 mm by 25 mm, so that each section measures 13 mm by 12.5 mm. The bar is suspended vertically and axially loaded in tension by a mass of 1.5 tonne. Calculate (a) the stress in each material, and (b) the extension.

The values of E for the steel and brass are 207 GN/m^2 and 110 GN/m^2 respectively.

$$\text{CSA of each bar} = 13 \times 12.5 \times 10^{-6} \,[\text{m}^2]$$
$$= 162.5 \times 10^{-6} \,[\text{m}^2]$$

$$\sigma_s A_s + \sigma_b A_b = F = mg$$

$$\sigma_s(162.5 \times 10^{-6}) + \sigma_b(162.5 \times 10^{-6}) = 1.5 \times 10^3 \times 9.81 \left[\frac{\text{kg m}}{\text{s}^2}\right]$$

$$= 14\,715\,\text{N} \qquad (1)$$

This has two unknown values, σ_s and σ_b, so we need another formula to solve it. Since the strains are equal in each bar,

$$\varepsilon_s = \varepsilon_b$$

Therefore $\dfrac{\sigma_s}{E_s} = \dfrac{\sigma_b}{E_b}$ and so $\dfrac{\sigma_s}{\sigma_b} = \dfrac{E_s}{E_b}$

$$\frac{\sigma_s}{\sigma_b} = \frac{207\,[\text{GN/m}^2]}{110\,[\text{GN/m}^2]} = 1.882$$

$$\sigma_s = 1.882\,\sigma_b \qquad (2)$$

Now we can substitute equation 2 into equation 1 for σ_s, leaving only one unknown value σ_b:

$$(1.882\,\sigma_b)(162.5 \times 10^{-6}) + \sigma_b(162.5 \times 10^{-6}) = 14\,715\,\text{N}$$

$$\sigma_b(305.825 + 162.5) \times 10^{-6} = 14\,715\,\text{N}$$

$$\sigma_b = \frac{14\,715\,[\text{N}]}{468.325 \times 16^{-6}\,[\text{m}^2]} = \mathbf{31.420 \times 10^6}\left[\frac{\text{N}}{\text{m}^2}\right]$$

$$\sigma_s = 1.882\,\sigma_b = 1.882 \times 31.420\left[\frac{\text{MN}}{\text{m}^2}\right] = \mathbf{59.133\ MN/m^2}$$

This value for σ_s could now be substituted into equation 1 to check the calculation if required.

To calculate the extension, we need to know the strain. As the strain for each material is the same, we can calculate this using values for either the steel or the brass.

$$\varepsilon = \frac{\sigma_s}{E_s} = \frac{59.133 \times 10^6\,[\text{N/m}^2]}{207 \times 10^9\,[\text{N/m}^2]} = 285.667 \times 10^{-6}$$

As $\varepsilon = \dfrac{x}{l}$ then $x = \varepsilon \times l$.

$$x = 285.667 \times 10^{-6} \times 0.5\,[\text{m}] = \mathbf{142.833 \times 10^{-6}\ m}$$

You can check this calculation using values from the brass rather than the steel.

$$\varepsilon = \frac{\sigma_b}{E_b} = \frac{31.420 \times 10^6}{110 \times 10^9}\left[\frac{\cancel{\text{N}}}{\cancel{\text{m}^2}}\right]\left[\frac{\cancel{\text{m}^2}}{\cancel{\text{N}}}\right] = 285.686$$

Shear stress and strain

So far we have looked at direct stresses (i.e. those stresses where the area being stressed lies at right angles to the direction of the line of force), and the materials either contract or extend. However, materials can deform in other ways.

Shear stress

The applied load may not consist of two forces acting in a straight line but instead of two forces that are equal and opposite and parallel that do not act in a straight line. Look at Figure 3.5.

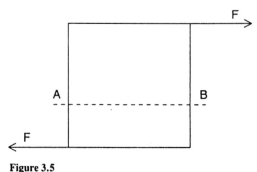

Figure 3.5

As these two forces F are not direct, there will be a tendency for one part of the material to slide over or shear from the other part. The cross-sectional area under shear stress is measured parallel to the line of action of the forces, e.g. at AB. The quantity symbol for shear stress is τ and the units are the same as direct stress: N/m^2.

$$\text{Average shear stress, } \tau = \frac{F}{A}\left[\frac{N}{m^2}\right]$$

❏ *Example 3.5*

A steel pivot bolt in a suspension arrangement has a maximum allowable shear stress of 60 MN/m^2 and a diameter of 25 mm. Find the maximum shear force that can be applied to the pivot.

$$\tau = \frac{F}{A} \quad \text{therefore } A = \tau \times A$$

$$\text{Area} = \frac{\pi d^2}{4} = \frac{\pi(25 \times 10^{-3})^2}{4} = \frac{\pi \times 25^2 \times 10^{-6}}{4}$$

$$\text{Maximum force, } F = 60 \times 10^6 \left[\frac{N}{m^2}\right] \times \frac{\pi(25)^2}{4}[mm^2] \times 10^{-6}\left[\frac{m^2}{mm^2}\right]$$

$$= 29.452 \times 10^3 \, N = \textbf{29.452 kN}$$

Shear strain

When shear stresses exist, the usual method of measuring the deformation, i.e. linear strain, cannot be applied and an alternative is required. This is called shear strain. Think of the block above subject to shear stress (Figure 3.6). As the load is applied, a deformation takes place as follows.

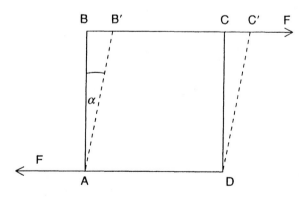

Figure 3.6

The square face ABCD is deformed to a parallelogram AB'C'D. Shear strain is then defined as the movement of the faces in the direction of the force divided by the distance between the faces.

$$\text{Shear strain} = \frac{BB'}{AB} = \tan\gamma = \gamma \text{ radians}$$

where γ is the angular distortion of the vertical faces. When an angle is very small $\tan \gamma \approx \gamma$ radians. (See page 104: $360° = 2\pi$ radians). Again there are no units of shear strain. You can think of strain as an angle of twist. Shearing is often associated with torques that tend to twist the material. Hooke's law still applies to shear stress and the relationship between shear stress and strain is called the modulus of rigidity and has the quantity symbol G.

$$G = \frac{\tau}{\gamma}$$

You can compare this to Young's modulus of elasticity. The values are different but tend to have the same sort of magnitude, e.g. for mild steel $E = 210\,[\text{GN/m}^2]$ and $G = 81\,[\text{GN/m}^2]$. Typical values of G for different engineering materials are shown in Table 3.2.

Table 3.2 Typical values for the modulus of rigidity

Material	$G[\text{GN/m}^2]$
Aluminium	25
Brass	37.5
Copper	50
Steel	80

☐ *Example 3.6*

A nickel steel gudgeon pin in an engine piston has an outside diameter of 27 mm and an inside diameter of 20 mm. The force on the piston is 8 kN. The modulus of rigidity for the steel is 90 GN/m². Calculate the stress and strain in the material.

$$\tau = \frac{F}{A}$$

$$A = \frac{\pi D^2}{4} - \frac{\pi d^2}{4} = \frac{\pi(D^2 - d^2)}{4}$$

$$\tau = \frac{F}{A} = \frac{8 \times 10^3\,[\text{N}]}{\left(\dfrac{\pi((27 \times 10^{-3})^2 - (20 \times 10^{-3})^2)}{4}\right)[\text{m}^2]}$$

$$= 30.960 \times 10^6 \left[\frac{\text{N}}{\text{m}^2}\right]$$

As $G = \dfrac{\tau}{\gamma}$ then $\gamma = \dfrac{\tau}{G}$

$$\gamma = \frac{\tau}{G} = \frac{30.960 \times 10^6\,[\text{N/m}^2]}{90 \times 10^9\,[\text{N/m}^2]} = 344 \times 10^{-6}$$

3.2 Bending of beams

Introduction

A beam is a length of rigid material, usually supported in a horizontal position carrying vertical loads. These vertical loads tend to do two things: bend the beam, and shear the beam. If the beam is bent, direct tensile and compressive stresses are caused, as you will see. When designing a beam, in order to calculate the material dimensions, both the direct stresses caused by the bending and the shear forces need to be considered. The theory of bending enables us to calculate the stresses so that the beam can be designed to carry these loads.

Many components of a vehicle have forces acting at right angles, such as axles and sections of the chassis, and can be considered as beams. Bending theory can then be applied. A vehicle chassis has to resist:

1. the weight of the vehicle, passengers and load
2. horizontal forces acting on the end of the chassis due to road shocks; these tend to distort the rectangular frame shape to a parallelogram
3. vertical forces acting upwards from the wheels due to road shocks; these tend to twist the frame and cause torsional distortion.

A chassis frame must have a cross-section that resists bending and distortion. The three usual types are channel, tubular or box, as shown in Figure 3.7.

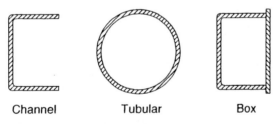

Channel   Tubular   Box

Figure 3.7 Chassis frame sections

The first part of this section looks at finding the magnitude of the bending moments and shear forces that a beam is subject to. The second part looks at the bending formula that is used to calculate the direct stresses and deflections caused by the bending moment. In all cases the forces acting on the beam are assumed to be in equilibrium; i.e. the algebraic sum of all horizontal and vertical forces acting on a beam is zero, and the algebraic sum of all moments of the forces is zero.

Shearing force and bending moments

Bending moment diagrams are used to find the bending moments that a beam is subject to along its length. Bending moments are simply moments of a force that tend to bend the beam. Shear force diagrams show the shear force that a beam is subject to along its length. In order to design a beam it is important to know the bending moment and shear force values at any section. You will sometimes find beams such as chassis members that vary in dimension along their length to stand the different forces they are subject to.

Shear forces

The definition of a shear force at any point along the length of a beam is the resultant vertical force of all the vertical forces acting to one side of the beam section. Look at Figure 3.8.

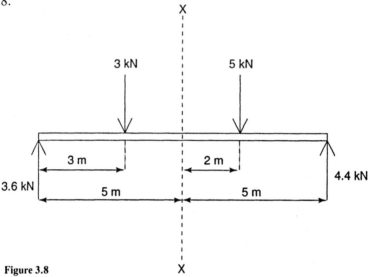

Figure 3.8

Take the section of the beam XX shown. For that portion of the beam to the right of the section,

resultant force $= 5\,[kN] \downarrow\, + 4.4\,[kN] \uparrow\, = 0.6\,kN \downarrow$.

For that portion of the beam to the left of the section,

resultant force $= 3.6\,[kN] \uparrow\, + 3\,[kN] \downarrow\, = 0.6\,kN \uparrow$.

The section XX is therefore subject to a vertical shearing force of 0.6 kN. This can be done for any section along the length of the beam.

A beam can be sheared in two ways: **positive shear** or **negative shear** (Figure 3.9). W is the resultant force acting on one side of the section.

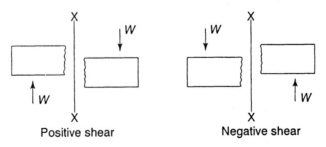

Figure 3.9

Bending moments

The definition of a bending moment at any section along the length of a beam, is the resultant moment about that section of all the forces acting to one side of that section. Go back to Figure 3.8. Consider the section XX again. For that portion of the beam to the right of the section,

resultant moment = $(5 \,[\text{kN}] \times 2 \,[\text{m}]) - (4.4 \,[\text{kN}] \times 5 \,[\text{m}]) = -12 \,\text{kNm}$

The negative sign indicates anti-clockwise. For that portion of the beam to the left of the section,

resultant moment = $(3.6 \,[\text{kN}] \times 5 \,[\text{m}]) - (3 \,[\text{kN} \times 2 \,[\text{m}]) = 12 \,\text{kNm}$

The section XX is therefore subject to a bending moment of 12 kNm. Two equal and opposite moments like this make a bending moment. The magnitude can be found by calculating the sum of the bending moments to either side of the section. A beam can be bent in two ways: **sagging** (positive) and **hogging** (negative) (Figure 3.10). M is the resultant moment due to the forces acting on one side of the section.

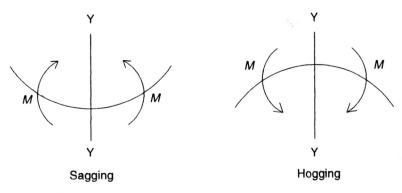

Sagging   Hogging

Figure 3.10

Moment of resistance

When the beam is bent, it stays in a state of equilibrium. When it is subject to a bending moment, then at any section there is an internal moment caused by internal forces to balance the bending moment applied at that section. This internal moment is called the **moment of resistance**.

Bending moment and shear force diagrams

There are two types of loads to consider: **concentrated loads** and **distributed loads**. Concentrated loads are where all the forces are assumed to be concentrated at specific points along the length of the beam. These are shown on sketches by an arrow at the precise point of application. Distributed loads are loads that are spread over a length of beam. The commonest type of distributed load is the weight of the beam itself. We assume that the centre of gravity of the distributed load acts at its

Distributed load

Figure 3.11 Distributed load

midpoint. Distributed loads are usually shown as a bumpy line on sketches (Figure 3.11).

A bending moment diagram is a type of graph. A horizontal base line is drawn to represent the length of the beam. Graphs are plotted above this base line. The vertical axis represents either shear force or bending moments. These diagrams are most easily explained by examples.

☐ *Example 3.7*

A beam which is simply supported at each end is 10 m long, and carries concentrated loads of 4 kN and 7 kN at 2 m and 6 m, respectively, from the left end. Neglect the weight of the beam. Draw the shearing force and bending moment diagrams.

The word 'respectively' just means that the values are taken in order, i.e. the 2 m refers to the 4 kN and the 6 m refers to the 7 kN. 'Simply supported' means that the beam is resting on two supports. These supports apply the upward reaction forces to resist the downward loads and the weight of the beam. Start off with a space diagram of the beam. See Figures 3.12–3.14.

The unknown forces are the two reaction forces R_A and R_B. These can be found by considering the moment first about one end and then about the other. Taking moments about the left end:

$$\text{clockwise moments} = \text{anti-clockwise moments}$$

$$(4\,[\text{kN}] \times 2\,[\text{m}]) + (7\,[\text{kN}] \times 6\,[\text{m}]) = R_B\,[\text{kN}] \times 10\,[\text{m}]$$

$$50\,[\text{kNm}] = R_B\,[\text{kN}] \times 10\,[\text{m}]$$

$$\therefore R_B = \frac{50\,[\text{kNm}]}{10\,[\text{m}]} = 5\,\text{kN}$$

(∴ means 'therefore', if you didn't know).

Taking moments about the right end:

$$R_A\,[\text{kN}] \times 10\,[\text{m}] = (4\,[\text{kN}] \times 8\,[\text{m}]) + (7\,[\text{kN}] \times 4\,[\text{m}])$$

$$R_A\,[\text{kN}] \times 10\,[\text{m}] = 60\,[\text{kNm}]$$

$$\therefore R_A = \frac{60\,[\text{kNm}]}{10\,[\text{m}]} = 6\,\text{kN}$$

You could also have calculated the second reaction force by:

$$\text{upward forces} = \text{downward forces}$$

$$5\,[\text{kN}] + R_A = 4\,[\text{kN}] + 7\,[\text{kN}]$$

$$\therefore R_A = 4\,[\text{kN}] + 7\,[\text{kN}] - 5\,[\text{kN}] = 6\,\text{kN}$$

Start the shear diagram with a base line to represent the length of the beam. The shear force line level changes suddenly when passing through a load point. The

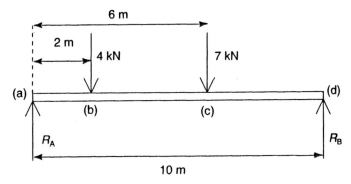

Figure 3.12 Space diagram

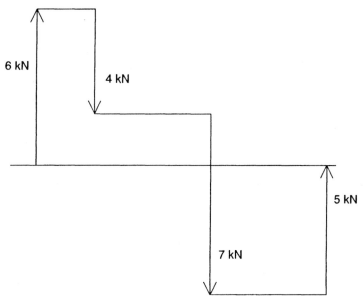

Figure 3.13 Shear force diagram

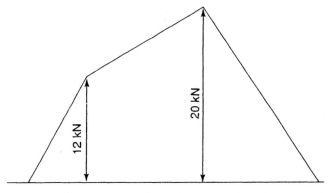

Figure 3.14 Bending moment diagram

change is equal to the load. At the left end of the beam at (a) there is an upward force of 6 kN, so draw a line vertically upwards, to scale, to represent the 6 kN. Moving to the right, there are no up or down forces until (b) at the downward 4 kN force. A line is drawn to scale to represent this vertically downwards at a scale distance of 2 m from the end. The tail of the 4 kN line is level with the head of the 6 kN and the line between them is horizontal. Further to the right there are no forces until (c) at the downward 7 kN force and so the line continues horizontally up to a scale distance of 6 m from the left end. Then the line is drawn vertically down to represent 7 kN; this line passes through the base line, and the line continues below it.

The last force of the system is the other reaction force R_B at (d) of 5 kN. This line drawn upwards closes the diagram.

If the shear force diagram does not close off a beam in equilibrium then you know a mistake has been made on the force calculations. The shear force line above the base line represents positive shear force and the line below the base line represents negative shear force.

Now that the shear force diagram is complete, the shear force at any point along the length of the beam can be determined. For example, at a point 3 m from the left end, the line is 2 kN above the base and so the shear force that the beam is subjected to here is +2 kN. The point where the shear force diagram passes through the base line is where the beam is subjected to the **maximum bending moment**.

When considering the shear force at a particular section, if you sum the forces to the left of a section making upward forces positive and downward forces negative, a positive value indicates a positive shear force and a negative value indicates a negative shear force. If you consider forces to the right of a section it's the other way around, with a negative value indicating a positive shear force.

Now draw the bending moment diagram. In order to draw the diagram, you first need to calculate the bending moment at various points along the beam. We will calculate bending moments at (a), (b), (c) and (d). Moments can be taken from either side. When moments are taken to the left of the points, positive values agree with our convention and indicate sagging. When working to the right of the points, negative values indicate sagging and positive values indicate hogging. For this example we will carry out the calculations to both sides to prove the point.

Bending moments to the left:

at (a) bending moments = 0
(There is no beam and therefore no moments to the left of (a), i.e. $R_A \times 0 = 0$)
at (b) $(6\,[kN] \times 2\,[m]) = 12\,[kN\,m]$
at (c) $(6\,[kN] \times 6\,[m]) - (4\,[kN] \times 4\,[m]) = 20\,[kN\,m]$
at (d) $(6\,[kN] \times 10\,[m]) - (4\,[kN] \times 8\,[m]) - (7\,[kN] \times 4\,[m])$
$= 0$

Bending moments to the right:
at (a) $(4\,[kN] \times 2\,[m]) + (7\,[kN] \times 6\,[m]) - (5\,[kN] \times 10\,[m])$
$= 0$
at (b) $(7\,[kN] \times 4\,[m]) - (5\,[kN] \times 8\,[m]) = -12\,kN\,m$
at (c) $-(5\,[kN] \times 4\,[m]) = -20\,kN\,m$
at (d) bending moments = 0

You would expect the beam to sag. This will not always be obvious though. Notice that the calculations taken from the left give positive values and all calculations taken from the right give negative values. If you stick to the convention when the problem is more complex, the signs will clearly indicate which way the beam bends. Draw positive values above the base line and negative values below, as with the shear force diagram. Select a suitable scale again so that the maximum bending moment will fit on the page.

You can join the points up with straight lines when all the loads are concentrated. From this diagram, the bending moment at any point along the beam can be directly measured. Halfway along the beam the bending moment is +18 kN m. You can check this by calculation:

$$(6\,[\text{kN}] \times 5\,[\text{m}]) - (4\,[\text{kN}] \times 3\,[\text{m}]) = +18\,\text{kN m}$$

Notice that the maximum bending moment occurs at the point where the shear force diagram crosses the base line. It is easier, if you have room, to draw the space diagram, the shear force diagram and the bending moment diagram all to the same scale and directly underneath each other.

After solving a few problems, you will realise that the shear force and bending moment diagrams follow standard patterns and you will be able to solve them quickly by finding certain values along the length of the beam and simply by joining them up with straight or curved lines. Distributed loads cause the joining lines of the bending moment diagram to be curved.

❏ *Example 3.8*

A cantilever is 10 m long and carries concentrated loads of 2 kN and 6 kN at the free end and at 4 m from the free end, respectively. Draw the shear force and bending moment diagrams. Neglect the weight of the beam. Determine the shear force and bending moment halfway along the beam.

A cantilever is a beam that is supported by being fixed into a wall at one end. The wall is usually drawn on the left of the beam. Start off with your space diagram (Figure 3.15).

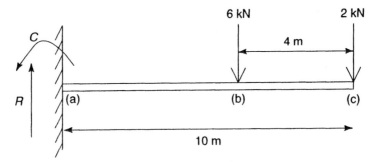

Figure 3.15

The wall exerts an upward reaction R. In this case this must equal 8 kN, as upward forces equal downward forces. Finding this value of R by equating moments would have been more complex. What is not at first obvious is that, for a cantilever, the wall exerts a fixing moment on the beam represented by C on the diagram. For this reason it is usually easier if bending moments are calculated to the right. Next draw the shear force diagram (Figure 3.16).

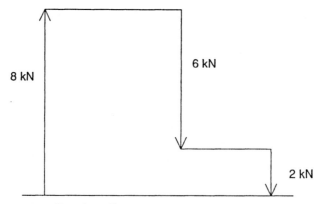

Figure 3.16 Shear force diagram

At the section halfway along the beam the shear force is 8 kN. The shear force diagram does not appear to cross the base line; it does in fact cross it at the immediate left and the maximum bending moment occurs at the wall. Next comes the bending moment calculations. Remember to take them to the right and so positive values will indicate negative bending moments, i.e. hogging.

at (a) $(2\,[\text{kN}] \times 10\,[\text{m}]) + (6\,[\text{kN}] \times 6\,[\text{m}]) = 56\,\text{kN m}$
at (b) $(2\,[\text{kN}] \times 4\,[\text{m}]) = 8\,\text{kN m}$
at (c) $= 0$

Draw the bending moment diagram below the base line as the positive values indicate negative bending moment and the beam is hogging (Figure 3.17).

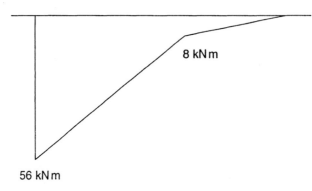

Figure 3.17 Bending moment diagram

The bending moment halfway along the beam is 16 kN m. You can check this by calculation:

$(2\,[\text{kN}] \times 5\,[\text{m}]) + (6\,[\text{kN}] \times 1\,[\text{m}]) = \mathbf{16\,kN\,m}$

Distributed loads

As was stated earlier, a distributed load is one which is spread over the length of the beam. The symbol for a distributed load is ω, which is the force per unit length of the

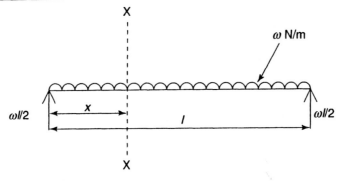

Figure 3.18

beam, and the unit is N/m. We will assume that all the distributed loads in this book are distributed evenly, rather than varying from point to point along the beam. Look at the beam in Figure 3.18; it has no loads placed on it, but its own weight is considered.

$$\text{Total load carried} = \omega \left[\frac{\text{N}}{\text{m}}\right] \times l\,[\text{m}]$$

As the beam is symmetrical, the reaction at the end supports are each ($\omega l/2$). The shear force diagram decreases from $+(\omega l/2)$ to 0 at the centre of the beam. The shear force then changes sign and the value increases to $-(\omega l/2)$. The straight line of the shear force slope is equal to the loading, ω.

$$\text{The shear force at a section XX} = +\frac{(\omega l)}{2} - \omega x$$

See Figure 3.19.

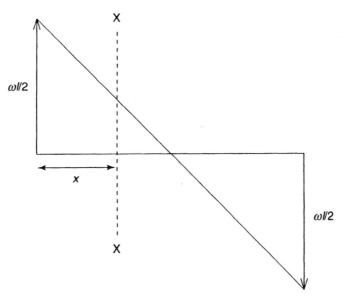

Figure 3.19 Shear force diagram of a simply supported distributed load beam

The bending moment at any section is found by treating the distributed load as an equivalent point load acting at the section's centre of gravity. So the moment of the distributed load to the left of section XX is

$$\omega x \times \frac{x}{2}$$

The bending moment at a section XX

$$= \frac{(\omega l)}{2} x - \omega x \frac{x}{2}$$

$$= \frac{\omega x}{2}(l - x)$$

Look at this last formula: when $x = 0$ or $x = l$, the bending moment will equal zero. The maximum bending moment will occur when $x = l/2$ (where the shear force diagram equals zero). If you substitute this into the formula the value will be $\omega l^2/8$. See Figure 3.20.

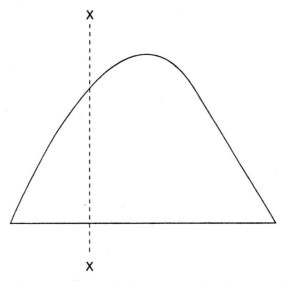

Figure 3.20 Bending moment diagram of a simply supported uniformly distributed loaded beam

$$\text{Bending moment} = \frac{\omega\left(\frac{l}{2}\right)}{2} \times \left(l - \frac{l}{2}\right)$$

$$= \frac{\omega\left(\frac{l}{2}\right)}{2} \times \frac{l}{2}$$

$$= \frac{\omega l^2}{8}$$

The bending moment diagram is a parabolic curve. A cantilever with a uniformly distributed load produces shear force and bending moment diagrams as shown in Figures 3.21, 3.22 and 3.23.

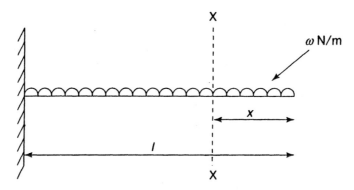

Figure 3.21

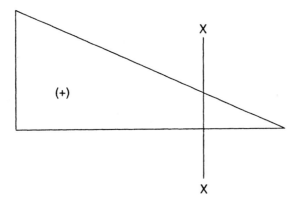

Figure 3.22 Shear force

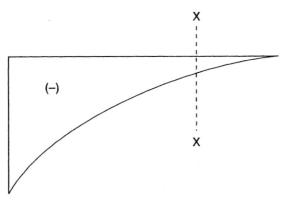

Figure 3.23 Bending moment

The equivalent concentrated load of the distributed load is ωl acting downwards at $l/2$ from the free end. So the reaction force on the beam by the wall is ωl upwards. At section XX, to the right, the shear force is $-\omega x$ which, according to our convention, is a positive shear force. The maximum shear force is at the wall. The bending moment at section XX is $+\omega x \times x/2 = +\omega x^2/2$, which, according to our convention, (as taken to the right) is hogging and so shown as negative on the bending moment diagram. The maximum bending moment is at the wall hogging and is equivalent to the fixing moment applied by the wall on the beam.

$$\text{Maximum bending moment} = \omega l \times \frac{l}{2} = \frac{\omega l^2}{2}$$

i.e. $\dfrac{\omega x^2}{2}$ when $x = l$

☐ Example 3.9

A simply supported beam is 10 m long and carries a load of 20 kN at 3 m from the right end. Assume that the weight of the beam is a uniformly distributed load of 5 kN/m. Draw the shear force and bending moment diagrams and determine the maximum bending moment. See Figure 3.24.

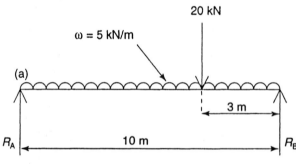

Figure 3.24

Remember to try to draw the space diagram, the shear force diagram and the bending moment diagram underneath each other and to the same length scale.

Total distributed weight $= 5\,[\text{kN/m}] \times 10\,[\text{m}] = 50\,\text{kN}$

Take moments about (a) to calculate R_B:

$$(20\,[\text{kN}] \times 7\,[\text{m}]) + (50\,[\text{kN}] \times 5\,[\text{m}]) = (R_B \times 10\,[\text{m}])$$

$$R_B = \frac{(140 + 250)\,[\text{kNm}]}{10\,[\text{m}]} = 39\,\text{kN}$$

$$(R_A \times 10\,[\text{m}]) = (20\,[\text{kN}] \times 3\,[\text{m}]) + (50\,[\text{kN}] \times 5\,[\text{m}])$$

$$R_A = \frac{(60 + 250)\,[\text{kNm}]}{10\,[\text{m}]} = 31\,\text{kN}$$

Check if upward forces equal downward forces:

$$31\,[kN] + 39\,[kN] = 20\,[kN] + 50\,[kN]$$

$$= 70\,kN \text{ in each case.}$$

The number of shear force and bending moment calculations you need to make will depend on your experience.

To draw the shear force diagram, first work out the slope of the distributed load according to the scale. In this case, the slope is 50 kN over 10 [m]. Look at Figure 3.25 and you will see how the distributed loads fit in with the direct loads. You can make calculations of the shear force at any point along the beam if need be.

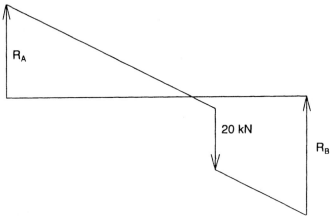

Figure 3.25

Shear force at:

0 m from (a)   = 31 kN ↑
2 m from (a) = 31 [kN] ↑ − (5 [kN/m] × 2 [m]) ↓ = 21 kN ↑
4 m from (a) = 31 [kN] ↑ − (5 [kN/m] × 4 [m]) ↓ = 11 kN ↑
6 m from (a) = 31 [kN] ↑ − (5 [kN/m] × 6 [m]) ↓ = 1 kN ↑

The points on the bending moment diagram will be joined by a parabolic curve. Calculate as many points as you feel necessary to draw the diagram accurately. Start off with the maximum bending moment value and measure the point of zero shear force down to the bending moment axis. This is at 6.2 m from the left end. The bending moment at the ends will be zero. Here are six points on the diagram.

Bending moment at:

0 m from (a) = 0

2 m from (a) = (31 [kN] × 2 [m]) − (5 [kN/m] × 2 [m] × 2 [m]/2) = 52 kN m

4 m from (a) = (31 [kN] × 4 [m]) − (5 [kN/m] × 4 [m] × 4 [m]/2) = 84 kN m

6.2 m from (a) = (31 [kN] × 6.2 [m]) − (5 [kN/m] × 6.2 [m] × 6.2 [m]/2)
                                                        = **96.1 kN m**

8 m from (a) = (31 [kN] × 8 [m]) − (20 [kN] × 1 [m])
$$- (5 [kN/m] × 8 [m] × 8 [m]/2) = 68 \, kN \, m$$

10 m from (a) = (31 [kN] × 10 [m]) − (20 [kN] × 3 [m])
$$- (5 [kN/m] × 10 [m] × 10 [m]/2) = 0$$

See Figure 3.26.

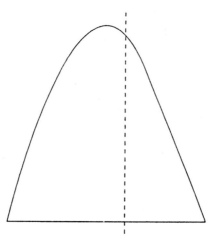

Figure 3.26 Bending moment

Tensile and compressive stress and the bending equation

Think of a rectangular sectioned beam which sags, as shown in Figure 3.27.

The top part of the beam will be compressed while the bottom part will be stretched. The greatest stress will occur at the outer edges of the beam, the top having compressive stresses and the bottom having tensile stresses. Towards the centre of the material the stress reduces uniformly. Where stresses change from compressive to tensile, there is zero direct stress. The axis along which there is no stress is called the neutral axis. For a rectangular section or any symmetrical section this is at the mid-depth. As the greatest stress is at the outer fibres of the beam then a rectangular section is not always an economical use of the material. An 'I' shaped sectioned beam is designed so that the top and bottom flanges resist the tensile and compressive stresses and all the cross-section resists the shearing forces. If the beam

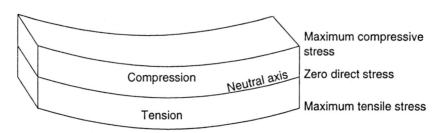

Figure 3.27

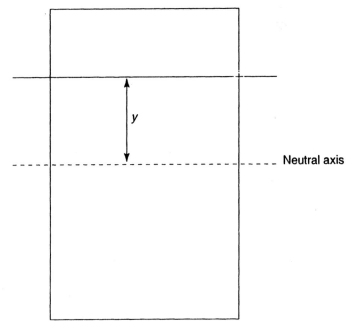

Figure 3.28 Cross-section of rectangular beam

bends the other way (hogging) then the tensile stresses are at the top of the material and the compressive stresses are at the bottom. The stress in the material is proportional to the distance from the neutral axis (NA). We'll call this distance y. See Figure 3.28. The stress at a distance of y from the NA is such that $\sigma_y \propto y$.

$$\therefore \sigma_y = y \times \text{constant, } K$$

$$\therefore \frac{\sigma}{y} = \text{constant, } K$$

For the section of the beam considered, there is a radius of curvature, R (Figure 3.29).

The constant in the first equation is equal to Young's modulus of elasticity for the beam material divided by the radius of curvature.

$$\frac{E}{R} = \text{constant, } K$$

So, we can write: $\dfrac{\sigma}{y} = \dfrac{E}{R}$

When a beam is loaded, the bending moment applied, represented by M, is resisted by an internal moment. This internal resisting moment is equal to the applied moment.

This can be written as: $M = \dfrac{E}{R} \times I$

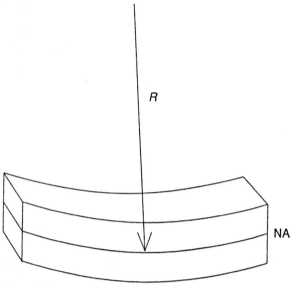

Figure 3.29

I is the second moment of area of the beam section about the neutral axis. More of this later. The whole equation can now be written as:

$$\frac{M}{I} = \frac{\sigma}{y} = \frac{E}{R}$$

The following symbols are used in the bending moment equations:

M = maximum bending moment (= resisting moment) [Nm]
I = second moment of area about the neutral axis [m⁴]

I = second moment of area about the neutral axis $[m^4]$
σ = stress at outer fibes of beam material $[N/m^2]$
y = distance from the neutral fibres to the outer fibres [m]
E = Young's modulus of elasticity for beam material $[N/m^2]$
R = Radius of curvature to the neutral axis [m].

Assumptions for bending moment equation

The bending moment equation is based on certain relationships between different quantities such as stress, elasticity, curvature and dimensions. For these relationships to be valid the following five assumptions are necessary:

1. The bending equation assumes that Hooke's law applies to each layer of the beam and Young's modulus of elasticity has the same value in compression as in tension.
2. The beam material is uniform throughout its volume.
3. Each cross-section of the beam is symmetrical about the plane of bending and the loads are applied to the beam in the plane of bending. If a horizontal beam supports vertical loads, then the plane of bending is vertical.
4. The beam must be initially straight and unstressed.
5. The resultant force perpendicular to any cross-section is zero.

Second moment of area

The amount of bending of a beam depends on the shape of the cross-section. This shape is expressed by the **second moment of area** about the neutral axis. The second moment of area is a measure of the distribution of an area relative to an axis. In the chapter on moments, the **first** moment of area is used to find the centroid of a plate.

$$\bar{x} = \frac{\Sigma \text{ moment of area}}{\Sigma \text{ area}}$$

Remember that the moment of area about an axis is the area multiplied by the distance from the centre of area to the axis. This is a difficult concept to visualise because it does not really physically exist: it is just a convenient measure to assist in calculations. The small circle in Figure 3.30 has an area of A.

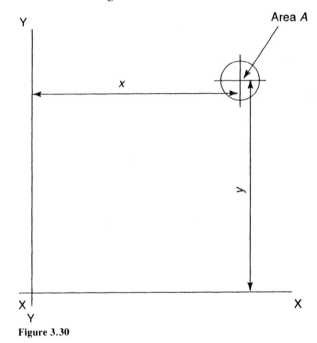

Figure 3.30

The first moment of area of A about $XX = Ay$
The second moment of area of A about $XX = Ay^2$
The first moment of area of A about $YY = Ax$
The second moment of area of A about $YY = Ax^2$

Working out the second moment of area of some larger shapes can be complicated. If any area is divided up into small bits, like the circle, and each bit multiplied by the distance to the axis squared, then all these products added up gives the second moment of area. As with centroids there are some standard cases for calculating the second moment of area about the neutral axis.

For a rectangular cross-section as shown in Figure 3.31

$$I_{NA} = \frac{bd^3}{12}$$

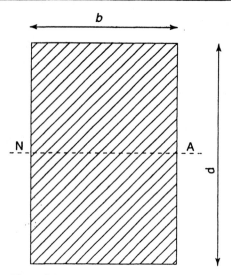

Figure 3.31

The second moment of area is represented by I and the axis about which moments are taken shown as a suffix, e.g. I_{XX} or I_{NA}

For a circular cross-section (Figure 3.32),

$$I_{NA} = \frac{\pi D^4}{64} = \frac{\pi r^4}{4}$$

For any of the shapes shown in Figures 3.33–3.35

$$I_{NA} = \frac{BD^3}{12} - \frac{bd^3}{12}$$

For a hollow circular section (Figure 3.36),

$$I_{NA} = \frac{\pi}{64}(D^4 - d^4) = \frac{\pi}{4}(R^4 - r^4)$$

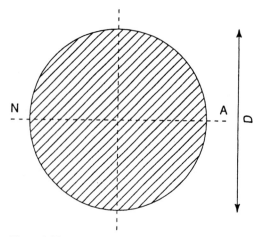

Figure 3.32

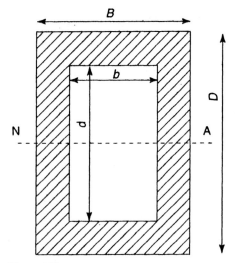

Figure 3.33

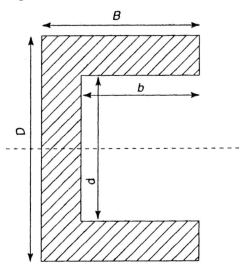

Figure 3.34

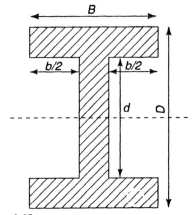

Figure 3.35

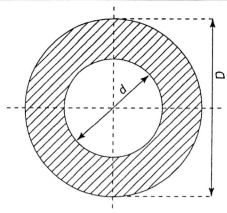

Figure 3.36

☐ *Example 3.10*

An engine is supported by lifting tackle hung from an overhead beam. The maximum bending moment M of the beam is 72 kNm. The second moment of area I is 176.68×10^{-6} m^4 and the beam is symmetrical. The depth of the beam is 276.8 mm. Calculate the maximum stress σ of the beam and the radius of curvature R. Young's modulus of elasticity = 208 GN/m^2.

Start by writing down the full form of the formula.

$$\frac{M}{I} = \frac{\sigma}{y} = \frac{E}{R}$$

From this, $\sigma = \dfrac{M}{I} \times y$

As the beam is symmetrical $y = \dfrac{276.88\,[\text{mm}]}{2} = 138.4$ mm.

$$\sigma = \frac{72 \times 10^3\,[\text{Nm}]}{176.68 \times 10^{-6}\,[\text{m}^4]} \times 138.4 \times 10^{-3}\,[\text{m}]$$

$$= \frac{72 \times 10^6}{176.68} \times 138.4 \left[\frac{\text{Nm}^2}{\text{m}^4}\right]$$

$$= 56.4 \times 10^6 \text{ N/m}^2$$

$$R = E \times \frac{I}{M}$$

$$= 208 \times 10^9 \left[\frac{\text{N}}{\text{m}^2}\right] \times \frac{176.68 \times 10^{-6}\,[\text{m}^4]}{72 \times 10^3\,[\text{Nm}]}$$

$$= 208 \times 10^9 \times \frac{176.68}{72} \times 10^{-9} \left[\frac{\cancel{\text{N}}\text{m}^{\cancel{4}}}{\text{m}^2\,\cancel{\text{N}}\text{m}}\right]$$

$$= \mathbf{510.409\ m} \quad \text{so there is not much of a bend}$$

□ *Example 3.11*

A beam has a rectangular cross-section 100 mm × 25 mm. It is subjected to a maximum bending moment of 625 Nm. Compare the maximum stress in the material when the beam is positioned with the widest edge down and with the narrowest edge down.

Start off with a diagram (Figure 3.37).

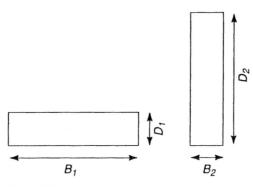

Figure 3.37

$$I_1 = \frac{0.1 \times (0.025)^3}{12} [\text{m}^4] = 0.1302 \times 10^{-6}\,\text{m}^4$$

$$y_1 = \frac{0.025}{2} [\text{m}] = 0.0125\,\text{m}$$

$$\sigma = \frac{M}{I} \times y = \frac{625\,[\text{Nm}]}{0.1302 \times 10^{-6}\,[\text{m}^4]} \times 0.0125\,[\text{m}]$$

$$= 60 \times 10^6\,[\text{N/m}^2]$$

$$I_1 = \frac{0.025 \times (0.1)^3}{12} [\text{m}^4] = 2.083 \times 10^{-6}\,\text{m}^4$$

$$y_2 = \frac{0.1}{2} [\text{m}] = 0.05\,\text{m}$$

$$\sigma = \frac{M}{I} \times y = \frac{625\,[\text{Nm}]}{2.083 \times 10^{-6}\,[\text{m}^4]} \times 0.05\,[\text{m}]$$

$$= 15.002 \times 10^6\,\text{N/m}^2$$

Notice that the stress is four times greater when the beam is positioned 'flat' rather than 'upright'.

To calculate the second moment of area of shapes where there is a part 'missing', then the second moment of area of the missing part can be subtracted from the second moment of area of the overall shape. The second moment of area of different shapes can be added or subtracted accordingly, provided they are relative to the same axis. This is probably easier to show with an example.

☐ *Example 3.12*

Look at Figure 3.38. Find the second moment of area about the neutral axis.

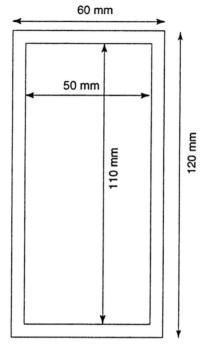

Figure 3.38

$$I_{NA} = \frac{BD^3}{12} - \frac{bd^3}{12} = \frac{60 \times 120^3}{12} - \frac{50 \times 110^3}{12} \, [\text{mm}^4]$$

$$= 8.640 \times 10^{-6} - 5.546 \times 10^{-6} \, [\text{m}^4]$$

$$= 3.094 \times 10^{-6} \, \text{m}^4$$

Parallel axis theorem

The parallel axis theorem is used for finding the second moment of area about an axis parallel to and other than the axis passing through the centroid, i.e. the neutral axis. This is also useful for finding the second moment of area of more complex shapes. Let I_{CG} be the second moment of an area A about an axis passing through its centroid. The distance to an axis parallel to axis CG is h. To find the second moment of the area about XX, I_{XX}, we can use the formula:

$$I_{XX} = I_{CG} + Ah^2$$

☐ *Example 3.13*

Find the second moment of area about the neutral axis of the cross-section shown in Figure 3.39.

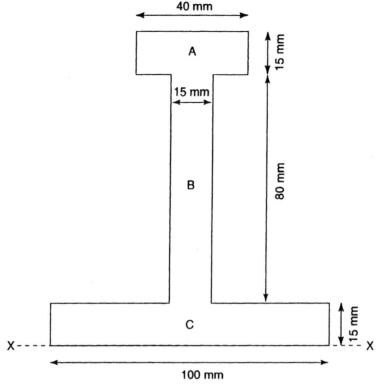

Figure 3.39

The first problem we have is that we do not know where the neutral axis is. As the neutral axis passes through the centroid, then we can use the formula:

$$\text{Distance of centroid from axis XX} = \frac{\Sigma \text{ 1st moment of area about XX}}{\Sigma \text{ area}}$$

Area

A $40 \times 10^{-3} \times 15 \times 10^{-3} \, [m^2] = 0.6 \times 10^{-3} \, m^2$

B $80 \times 10^{-3} \times 15 \times 10^{-3} \, [m^2] = 1.2 \times 10^{-3} \, m^2$

C $100 \times 10^{-3} \times 15 \times 10^{-3} \, [m^2] = \underline{1.5 \times 10^{-3} \, m^2}$

$3.3 \times 10^{-3} \, m^2$

First moment of area

A $0.6 \times 10^{-3} \, [m^2] \times 102.5 \times 10^{-3} \, [m] = 61.500 \times 10^{-6} \, m^3$

B $1.2 \times 10^{-3} \, [m^2] \times 55 \times 10^{-3} \, [m] = 66.000 \times 10^{-6} \, m^3$

C $1.5 \times 10^{-3} \, [m^2] \times \dfrac{15}{2} \times 10^{-3} \, [m] = \underline{11.250 \times 10^{-6} \, m^3}$

$138.750 \times 10^{-6} \, m^3$

Distance of centroid from axis $XX = \dfrac{138.750 \times 10^{-6}\,[\text{m}^3]}{3.3 \times 10^{-3}\,[\text{m}^2]}$

$$= 42.045 \times 10^{-3}\,\text{m}$$

Hence the centroid lies at a distance of 42.045 mm from XX.

Hence the neutral axis lies 42.045 mm from XX, and we can calculate the second moment of the individual areas relative to this. These second moment of areas can then be added up, as they are taken from the same axis.

$A \quad I_{NA(A)} = \dfrac{bd^3}{12} + Ah^2$

$$= \dfrac{(0.04 \times 0{,}015^3)}{12}\,[\text{m}^4] + 0.6 \times 10^{-3}\,[\text{m}^2]$$
$$\times ((102.5 - 42.045) \times 10^{-3})\,[\text{m}^2]$$
$$= 11.25 \times 10^{-9}\,[\text{m}^4] + 2192.884 \times 10^{-9}\,[\text{m}^4]$$
$$= 2204.134 \times 10^{-9}\,\text{m}^4$$

$B \quad I_{NA(A)} = \dfrac{bd^3}{12} + Ah^2$

$$= \dfrac{(0.15 \times 0.08^3)}{12}\,[\text{m}^4] + 1.2 \times 10^{-3}\,[\text{m}^2]$$
$$\times ((55 - 42.045) \times 10^{-3})\,[\text{m}^2]$$
$$= 640 \times 10^{-9}\,[\text{m}^4] + 201.398 \times 10^{-9}\,[\text{m}^4]$$
$$= 841.398 \times 10^{-9}\,\text{m}^4$$

$C \quad I_{NA(A)} = \dfrac{bd^3}{12} + Ah^2$

$$= \dfrac{(0.1 \times 0.015^3)}{12}\,[\text{m}^4] + 15 \times 10^{-3}\,[\text{m}^2]$$
$$\times ((42.045 - 7.5) \times 10^{-3})\,[\text{m}^2]$$
$$= 28.125 \times 10^{-9}\,[\text{m}^4] + 1790.036 \times 10^{-9}\,[\text{m}^4]$$
$$= 1818.161 \times 10^{-9}\,\text{m}^4$$

Total, $I_{NA} = A + B + C$

$$= 2204.134 \times 10^{-9}\,[\text{m}^4]$$
$$+ 841.398 \times 10^{-9}\,[\text{m}^4] + 1818.161 \times 10^{-9}\,[\text{m}^4]$$
$$= \mathbf{4863.693 \times 10^{-9}\,\text{m}^4}$$

Remember when solving the beam problems at the end of the chapter not to get confused between shear stress and direct stress. The bending moment and shear force can be found at any point along the beam with bending moment and shear force diagrams. The maximum direct stress and curvature can be calculated from the maximum bending moment. Stick to the convention for shear force and bending moments, as often the sign of the answer will not be obvious.

3.3 Torsion

When the moment of a force is applied to a shaft, it will tend to turn or twist. This is referred to as a turning moment or torque. Torsion is the twisting deformation produced when a torque is applied to a shaft. When a shaft twists, if any thin circular cross-section is considered it will rotate slightly relative to the next section. The material therefore undergoes shear strain and shear stress, as thin circular cross-sections of material tend to slide over each other. This is an important subject in motor vehicle science: consider the number of shafts in a motor vehicle, from the crankshaft, through the gear box and the drive shafts to the axles. There is a similar equation for the torsion of shafts as there is for the bending of beams. This equation relates the applied torque to the shear stress and twisting distortion of the shaft in a similar manner that the bending equation relates the bending moment to direct stress and bending distortion of the beam. Torque is the twisting moment applied to the shaft and is a product of applied force and the distance of the applied force to the central axis of the shaft. The symbol for torque is T and its units are Nm (see 'moments'). When a shaft is transmitting power, it is subject to a torque at the input end from the drive such as an engine and gear box, and a resisting torque at the output from the machinery being driven, such as axle and wheels.

Angle of twist

When a torque is applied to a shaft, it twists. There is a relative movement between adjacent cross-sections of the shaft. If a straight line is drawn down the shaft parallel to the axis, when the torque is applied this deforms to a helix. The relative angular twist between the different cross-sections is called the angle of distortion, α. See Figure 3.40.

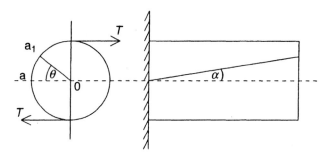

Figure 3.40

A torque T is applied to the shaft whilst one end is kept fixed. A point a on the end section is displaced to a position a_1. The angle of twist at the shaft face is $\theta = a\hat{o}a_1$. The angle θ is measured in radians. If a length of one radius r is laid along the circumference of the circle, then the angle, θ, between each end of this length and the circle centre is one radian. As the circumference of a circle is equal to πd or $2\pi r$, then in 360° there are 2π radians. See Figure 3.41.

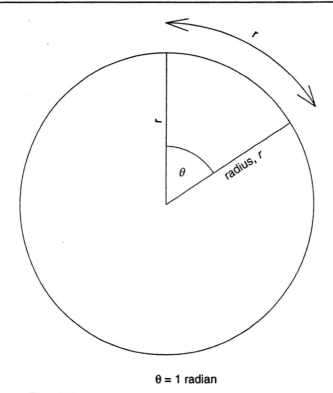

$\theta = 1$ radian

Figure 3.41

To convert from radians to degrees use the following relationship:

$360° = 2\pi$ radians

So, for the unity bracket, $1 = \dfrac{2\pi}{360}\left[\dfrac{\text{radians}}{°}\right] = \dfrac{360}{2\pi}\left[\dfrac{°}{\text{radians}}\right]$

Radians are normally indicated by a small letter r after the angle value θ^r, although radians do not really have a unit. In engineering you can assume that most angles are measured in radians rather than degrees. This makes calculations and units much simpler to deal with than degrees, as you will see.

Shear stress

The value of the shear stress is proportional to the radius, zero shear stress occurring at the shaft section centre and maximum shear stress occurring at the outer edge. Mathematically this is:

$\tau = r \times$ constant

If this constant is represented by the letter K then:

$\dfrac{\tau}{r} = K$

Torsion equation

The above constant is used as the basis of the torsion equation. The full equation is as follows:

$$\frac{T}{J} = \frac{\tau}{r} = \frac{G\theta}{l}$$

You will be probably be familiar with most of these symbols except for the J.

T = Twisting moment or torque   [Nm]
J = Polar second moment of area   [m⁴]
τ = Maximum shear stress at outer fibres   [N/m²]
r = Radius to outer fibres   [m]
G = Modulus of rigidity for shaft material   [N/m²]
θ = Angle of twist   [radians]
l = Length of shaft under torsion   [m]

For a shaft, the radius to the outer fibres is simply the radius of the shaft, and this is where the maximum shear stress occurs.

Polar second moment of area

In bending theory for beams, the second moment of area is taken from the neutral axis. The resistance to twisting depends on the polar second moment of area J. The polar second moment of area is the second moment of area of the shaft cross-section about the central point rather than an axis. When you are dealing with shafts there are usually only two cases: either a solid circular shaft or a hollow circular shaft with an annular (doughnut shaped) cross-section.

For a solid circular shaft,

$$J = \frac{\pi D^4}{32} \quad \text{where } D \text{ is the diameter}$$

For a hollow circular shaft,

$$J = \frac{\pi (D^4 - d^4)}{32} \quad \text{where } d \text{ is the inner diameter}$$

☐ Example 3.14

When a torque of 100 Nm is applied to a solid shaft, the angle of twist at the shaft face is 1°. The shaft is 1 m long and for the shaft material, $G = 83$ GN/m². Calculate the diameter of the shaft.

As we are given T, θ, G and l then we can use the expression:

$$\frac{T}{J} = \frac{G\theta}{l}$$

to find J and then from that the diameter.

$$1° = 1° \times \frac{2\pi}{360} \left[\frac{rads}{°}\right] = 17.453 \times 10^{-3} \, radians$$

$$J = T \times \frac{l}{G\theta}$$

$$= 100 \, [Nm] \times \frac{1}{83 \times 10^9 \times 17.453 \times 10^{-3}} \, [m]\left[\frac{m^2}{N}\right]$$

$$= 69.032 \times 10^{-9} \, m^4$$

For a solid shaft, $J = \dfrac{\pi D^4}{32}$

$$\text{so } D = \sqrt[4]{\left(\frac{J \times 32}{\pi}\right)} = \sqrt[4]{\left(\frac{69.032 \times 10^{-9} \times 32}{\pi}\right)} \sqrt[4]{[m^4]}$$

$$= 28.958 \times 10^{-3} \, m = \mathbf{28.96 \, mm}$$

Notice how the units are subject to the $\sqrt[4]{}$ as well as the numbers.

☐ *Example 3.15*

If a drive shaft 2 m long is to be subjected to a torque of 3.3 kN and the maximum shearing stress must not exceed 48 MN/m², calculate the solid shaft diameter required to satisfy these conditions. $G = 80$ GN/m². If we use:

$$\frac{T}{J} = \frac{\tau}{r}$$

it appears as if we have two unknowns, J and r. We can break this formula down further:

$$T \times r = \tau \times J \Rightarrow T \times \frac{D}{2} = \tau \times \frac{\pi D^4}{32}$$

The diameter D on the left can be cancelled with a diameter on the right.

$$\frac{T}{2} = \tau \times \frac{\pi D^3}{32} \quad \text{therefore} \quad D = \sqrt[3]{\frac{T \times 32}{2 \times \tau \times \pi}}$$

$$= \sqrt[3]{\frac{3.3 \times 10^3 \times 32}{2 \times 48 \times 10^6 \times \pi}} \sqrt[3]{[Nm]\left[\frac{m^2}{N}\right]} = \mathbf{70.482 \times 10^{-3} \, m}$$

A diameter of 70.482 mm would satisfy these conditions of a maximum shearing stress of 48 MN/m². If there were further conditions, such as a maximum angle of twist, then further calculations would be required. Assume that the maximum angle of twist permissible for the above shaft is 1°.

$$1° = 17.453 \times 10^{-3} \, radians \qquad \text{(see Example 3.14)}$$

using $\dfrac{T}{J} = \dfrac{G\theta}{l}$

then $J = T \times \dfrac{l}{G\theta}$

$$= \dfrac{3.3 \times 10^3 \, [\text{Nm}] \times 2 \, [\text{m}]}{80 \times 10^9 \left[\dfrac{\text{N}}{\text{m}^2}\right] \times 17.453 \times 10^{-3}}$$

$$= 4.727 \times 10^{-6} \, \text{m}^4$$

$$J = \dfrac{\pi D^4}{32},$$

therefore $D = \sqrt[4]{\left(\dfrac{J \times 32}{\pi}\right)}$

$$= \sqrt[4]{\left(\dfrac{4.727 \times 10^{-6} \, [\text{m}^4] \times 32}{\pi}\right)}$$

$$= 83.300 \times 10^{-3} \, \text{m} = \textbf{83.3 mm}$$

In order to satisfy the maximum shearing stress of $48\,\text{MN/m}^2$ and the maximum angle of twist of $1°$, the larger diameter would be required: 83.3 mm.

Torsional resilience

Torsional resilience is the elastic strain energy stored in the shaft when work is done to twist the shaft. As with direct resilience (see 'stress and strain') this energy is only stored if the strain is below the elastic limit. If a torque is gradually applied from zero to T, then the average torque applied is

$$\dfrac{0 + T}{2} = \dfrac{T}{2}$$

The work done to twist the shaft is therefore equal to

$$\dfrac{T}{2} \times \theta^r$$

Again, as with direct resilience, this can be expressed in terms of stress.

From $\dfrac{T}{J} = \dfrac{\tau}{r}$

$$\Rightarrow T = \dfrac{\tau J}{r}$$

For a solid shaft, $J = \dfrac{\pi D^4}{32}$, and $r = \dfrac{D}{2}$

then $T = \tau \times \dfrac{\pi D^4}{32} \times \dfrac{2}{D}$

$\qquad = \dfrac{\tau \pi D^3}{16}$

From $\dfrac{\tau}{r} = \dfrac{G\theta}{l}$

$\qquad \theta = \dfrac{\tau l}{rG} = \dfrac{\tau l 2}{GD}$

Using the formula for torsional resilience:

$\dfrac{T\theta}{2} = \dfrac{1}{2} \times \dfrac{\tau \pi D^3}{16} \times \dfrac{\tau l 2}{GD}$

$\qquad = \dfrac{\tau^2 \pi D^2 l}{16G}$

The volume, $V = \dfrac{\pi D^2}{4} \times l$

Therefore $\dfrac{T\theta}{2} = \dfrac{\tau^2 V}{4G}$

❑ *Example 3.16*

In the last example, assume a diameter of 83.3 mm is selected. Calculate the torsional resilience.

$\dfrac{T}{J} = \dfrac{\tau}{r}$

so, $\tau = \dfrac{Tr}{J}$

$= \dfrac{3.3 \times 10^3\,[\text{Nm}] \times 83.3/2 \times 10^{-3}\,[\text{m}]}{\left(\dfrac{\pi(83.3 \times 10^{-3})^4}{32}\right)[\text{m}^4]}$

$= \dfrac{3.3 \times 41.65 \times 10^{12}}{4\,726\,937.9}\left[\dfrac{\text{N}}{\text{m}^2}\right]$

$= 29.077 \times 10^6\ \text{N/m}^2$

Torsional resilience $= \dfrac{\tau^2 V}{4G}$

$= \dfrac{(29.077 \times 10^6)^2\,[\text{N/m}^2]^2 \times \left(\dfrac{\pi(83.3 \times 10^{-3})^2}{4}\right)[\text{m}^2] \times 2\,[\text{m}]}{4 \times 80 \times 10^9\,[\text{N/m}^2]}$

$$= \frac{845.472 \times 5449.792 \times 2 \times 10^{-3}}{4 \times 80} \left[\frac{N^2}{m^4}\right]\left[\frac{m^2}{N}\right][m^3]$$

$$= 28.798 \text{ Nm}$$

As we know the angle of twist in this case, 17.453×10^{-3} radians, then the problem could have actually been solved with just the basic formula for torsional resilience:

$$\text{torsional resilience} = \frac{T\theta}{2}$$

$$= \frac{3.3 \times 10^3 \,[\text{Nm}] \times 17.453^{\text{r}}}{2}$$

$$= 28.798 \text{ Nm}$$

Problems 3

1. Calculate the average shear stress in a solid gudgeon pin of 30 mm diameter. The load on the pin is 22 kN.
2. A round bar has a length 270 mm and diameter 15 mm. A tensile load is gradually applied to the bar. A load of 25 kN causes an extension of 0.2 mm. Find the value of Young's modulus of elasticity for the bar material.
3. A push rod has a diameter of 5 mm. At a certain point in the operation of the valve, the force on the rod is 2.5 kN. Calculate the compressive stress in the rod.
4. A piston has a force of 4 kN acting on its crown at top dead centre. The cross-sectional area of the connecting rod is 180 mm². Calculate the compressive stress in the material.
5. What is meant by the term resilience?
6. A steel bar has a diameter of 50 mm and is 1 m long. The bar supports a tensile load of 50 kN. Calculate the resilience of the bar, if $E = 200 \text{ GN/m}^2$.
7. A cast iron cylinder has an outside diameter of 120 mm and an inside diameter of 100 mm, and supports a load of 100 kN. Find the compressive stress in the material.
8. A cantilever is 1.5 m long and supports concentrated loads of 10 kN at the free end and 15 kN at 0.75 m from the wall. Draw the shear force and the bending moment diagrams to scale and measure the shear force and bending moment 0.7 m from the wall.
9. A beam of length 7 m is simply supported at each end and carries a concentrated load at 3 m from one end. If the bending moment is 50 kN m at the section where the load is concentrated, find the magnitude of the load and the reaction forces.
10. A beam carries a uniformly distributed load over its entire length. The distributed load plus the weight of the beam equal 65 kN. The beam is 3 m long and simply supported at each end. Draw the shear force and bending moment diagrams and find the maximum bending moment.
11. Two cables, one made of steel and one made of copper, are of equal length and used together to support a mass of 30 kg. The diameter of the steel cable is 1.2 mm and the diameter of the copper cable is 2 mm. Calculate the load taken by each wire. For steel, $E = 200 \text{ GN/m}^2$, and for copper, $E = 98 \text{ GN/m}^2$.

12. A hole of diameter 5 mm needs to be punched in a plate 2.5 mm thick. The ultimate shear stress of the plate material is $350 \, \text{MN/m}^2$. Calculate the force required. *Hint*: The area resisting shear is equal to the hole circumference multiplied by the thickness.

13. Calculate the maximum thickness of steel plate that can be cut by a guillotine if the ultimate shear stress of the plate is $300 \, \text{MN/m}^2$ and the maximum guillotine force is 150 kN. The plate is 0.8 m wide.

14. How many 10 mm diameter rivets are required for a single lap joint which is to support a load of 100 kN? The safe working shear stress of the rivet material is $75 \, \text{MN/m}^2$.

15. A beam is 200 mm deep by 100 mm wide and subject to a maximum bending moment of 300 kNm. What is the maximum stress in the beam? The value of E for the beam material is $200 \, \text{GN/m}^2$.

16. A steel rectangular bar is 12 mm thick and is bent to an arc of a circle until the steel just yields at the top surface. The yield stress of the material is $300 \, \text{MN/m}^2$ and $E = 200 \, \text{GN/m}^2$. Find the radius of curvature.

17. A solid steel shaft has a diameter of 300 mm and twists through 1° over a length of 5 m when rotating at 100 rev/min. The value of $G = 85 \, \text{GN/m}^2$. Calculate the torque imposed on the shaft.

18. A torque of 100 Nm is applied to a shaft 2 m long. The angle of twist is 0.5°. The value of $G = 85 \, \text{GN/m}^2$. Calculate the diameter of the shaft.

19. A beam of length 5 m is simply supported at the ends. It carries concentrated loads of 100 N, 200 N and 300 N at distances of 1 m, 2.5 m and 4 m from one end, respectively. Calculate the reaction forces.

20. Find the twisting moment of a solid shaft of diameter 100 mm when the angle of twist is 0.75° over a length of 2 m. For the shaft material, $G = 100 \, \text{GN/m}^2$.

21. A hollow shaft has an outside diameter of 350 mm and an inside diameter of 250 mm. It transmits a torque of 400 kNm. Calculate the shear stress if $G = 90 \, \text{GN/m}^2$.

4 Motion

This chapter looks at the way a body moves. Newton's laws of motion are considered throughout.

4.1 Linear motion

The main quantities of interest to us in motion are velocity, acceleration, displacement and time. As these are all vector quantities, then many problems can be solved by simple vector diagrams.

Displacement

If a body moves from one point to another then it is **displaced**. The **displacement** is a vector quantity and the magnitude and the direction must be stated. The unit of displacement is the metre, m. The symbol for displacement is S.

Velocity

The **velocity** of a body is the rate of change of displacement, i.e. how far it moves in a unit of time. The symbol of velocity is v and the unit is metres per second (m/s). If a body moves 10 m in one second, the next second moves another 10 m, and the next second moves another 10 m, then in three seconds the body moves 30 m. The average velocity is 10 m/s. If the velocity is constant, then the displacement each second remains constant. If the velocity is not constant then the displacement each second is different. dS/dt may be written to mean the velocity at any particular instant in time. Velocity is a vector quantity and so the magnitude and the direction must be stated. The term **speed** refers to the magnitude part of velocity as speed is a scalar quantity. For example, the speed of a car could be 20 m/s but the velocity would be 20 m/s due east.

Acceleration

Any change in the velocity of a body will result in an acceleration. **Acceleration** to velocity is what velocity is to displacement. Acceleration is the rate of change of velocity. The symbol for acceleration is a and the units are m/s^2. If the velocity of a body one second is 10 m/s, the next second is 15 m/s, and the next second is 20 m/s, then the acceleration is 5 m/s per second, i.e. $+5$ m/s^2. A negative acceleration indicates that the body is slowing down. Linear retardation is negative acceleration. d^2S/dt^2 may be written to mean the acceleration at any particular instant in time. Acceleration is also a vector quantity and so direction needs to be stated with the magnitude.

Newton's laws of motion

Newton's first and second laws of motion have already been mentioned briefly in the chapter on mass. All three laws of motion are as follows.

Newton's first law of motion

A body continues in its state of rest, or of uniform motion, in a straight line unless it is acted upon by an external force.

A body continuing in a state of rest seems familiar enough. If a car stands on a level road and no force acts on the vehicle, we would not expect it to go anywhere. The body continuing in a state of uniform motion may seem a bit odd though, as on Earth we are used to moving things slowing down and finally stopping when the driving force is removed; this happens because of resistive frictional forces. Without friction, such as in space, a body will continue in its state of uniform motion in a straight line until an external force causes it to do otherwise. If a vehicle travels in a straight line at a constant velocity, then the driving force that the engine provides equals the resistive forces of friction and air resistance, and the resultant force is zero. For the car to accelerate, the driving force must exceed the resistive forces, so the resultant force acts in the direction of motion. If the driving force is less than the resistive forces, then the resultant force opposes motion and the vehicle slows down. It is important to be aware of this law of motion when considering what forces will affect the motion of a body.

Newton's second law of motion

The external force acting on a body is proportional to the rate of change of momentum of the body, the change of momentum being in the direction in which the force acts.

In the Chapter 2 on mass we looked at the relationship between force, mass and acceleration, i.e. $F = m \times a$, which enables us to measure a force. This, however, is just one way of changing the momentum of a body.

Momentum = mass × velocity   [kg m/s]

The second law is concerned with the rate of change of momentum:

rate of change of momentum $= \dfrac{mv - mu}{t}$

The law mathematically can be written as:

$$F \propto \frac{mv - mu}{t}$$

This force can change by the mass altering or the velocity altering, and, because velocity is a vector quantity, then this force can alter either through change in direction or change in magnitude. If the mass of the body does not change, then the law can be written as:

$$F \propto \frac{m(v - u)}{t}$$

The rate of change of velocity, $(v - u)/t$, is acceleration a.

So $F \propto m \times a$ and $F = m \times a \times$ constant.

The SI system is arranged so that the constant value is unity, 1, leaving $F = m \times a$.

The velocity of a body can change in magnitude through the acceleration or deceleration of the body. Its velocity can change in direction, for instance it could move round in a circle at a constant speed, i.e. the momentum changes because the direction of the velocity changes.

Example 4.1

A body moves with a velocity of 10 m/s due east. A force acts upon the body for two seconds and the body's velocity changes to 10 m/s due south. Find the change in velocity and the average acceleration. See Figures 4.1 and 4.2.

Change in velocity = final velocity − initial velocity

To subtract a vector quantity on a vector diagram, reverse the vector direction and add:

The change in velocity = **14.142 m/s**

$$\text{Acceleration} = \frac{\text{change in velocity}}{\text{time}}$$

$$= \frac{14.142 \,[\text{m/s}]}{2 \,[\text{s}]} = 7.071 \text{ m/s}^2$$

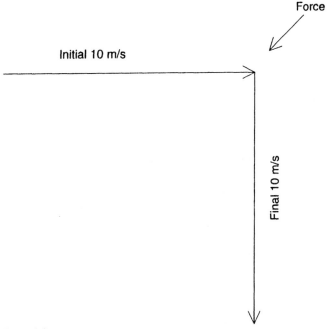

Figure 4.1

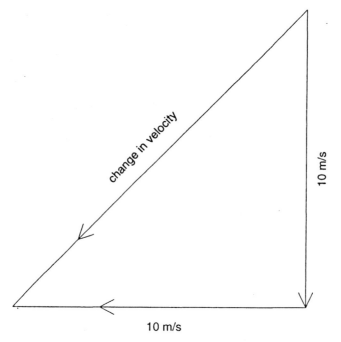

Figure 4.2

A change in momentum is always accompanied by a force. The value of this force will depend upon the rate at which the momentum is changed. The force is greater when the momentum is changed quickly.

A measure of a body's resistance to change in motion is called **inertia** and is due to its mass.

Newton's third law of motion

To every action there is always an equal and opposite reaction.

An example of this is someone jumping from a small boat to a river bank. The person will usually fall in the water, because as they leap forward, the boat moves in the opposite direction backwards.

Equations of motion

There are some basic equations of motion that describe the relationship between displacement, velocity, acceleration and time. Consider a body that moves between two points. The distance (in a straight line) between the two points is S metres. Let the time taken to pass between the two points be t seconds. The body does not move with a constant velocity though: it moves with a steady increase in velocity, i.e. a constant acceleration. The velocity as the object passes the first point is u [m/s] and the velocity as the object passes the second point is v [m/s].

$$\text{Average velocity} = \frac{u + v}{2}$$

which is the velocity at the mid-point in time.

We know also that: average velocity $= \dfrac{\text{displacement}}{\text{time}}$

$\therefore$ displacement, $S = \dfrac{u + v}{2} \times t$ (1)

Acceleration is the rate of change of velocity:

$$\text{acceleration} = \dfrac{\text{change in velocity}}{\text{time taken}}$$

or, $a = \dfrac{v - u}{t}$

Rearranging this to find v:

$\therefore v = u + at$ (2)

This gives us a formula for final velocity in terms of acceleration, time and initial velocity. Now, if we substitute for velocity from equation 2 into equation 1 we obtain an equation for displacement in terms of acceleration, time and initial velocity.

$$S = \dfrac{u + v}{2} \times t$$

$$= \dfrac{u + at + u}{2} \times t$$

$$= \dfrac{2u + at}{2} \times t$$

$$\Rightarrow S = ut + \dfrac{1}{2} at^2 \qquad (3)$$

If we rearrange equation 2, $t = (v - u)/a$.

We can now substitute for time into equation 1 to obtain an equation for velocity in terms of acceleration, displacement and initial velocity:

$$S = \dfrac{u + v}{2} \times t$$

$$\Rightarrow = \dfrac{u + v}{2} \times \dfrac{v - u}{a}$$

$$\Rightarrow = \dfrac{v^2 - u^2}{2a}$$

rearrange to $v^2 = u^2 + 2aS$ (4)

❏ *Example 4.2*

A car travelling down the road increases in velocity from 8.9 m/s to 17.9 m/s, (about 20 mile/h to 40 mile/h). This takes 20 s. Calculate the distance travelled during this period and the acceleration of the car.

To find out the distance we can use equation 1:

$$S = \frac{u+v}{2} \times t$$

$$= \frac{8.9 + 17.9}{2} \left[\frac{m}{s}\right] \times 20 \,[s] = \textbf{268 m}$$

We have enough information to use any of the other formulae to calculate acceleration. Equation 2 looks the simplest.

$$v = u + at$$

rearranging gives $a = \dfrac{v-u}{t} = \dfrac{17.9 - 8.9}{20} \left[\dfrac{m}{s}\right]\left[\dfrac{1}{s}\right]$

$$= \textbf{0.45 m/s}^2$$

That is an increase in velocity of 0.45 m/s every second.

❏ Example 4.3

A vehicle reduces its velocity to a standstill over a displacement of 500 m with a retardation of 10 m/s². Calculate its initial velocity.

We know that $S = 500$ m, $a = -10$ m/s² and $v = 0$. We need to find the initial velocity, u.

Use equation 4: $v^2 = u^2 + 2aS$

rearrange, to give $u^2 = v^2 - 2aS$

$$u = \sqrt{v^2 - 2aS}$$

$$= \sqrt{0 - (2 \times -10 \times 500)} \left[\sqrt{\frac{m}{s^2}\, m}\right]$$

$$= \textbf{100 m/s}$$

Velocity–time diagrams

A velocity–time diagram is a graph of velocity against time. Time is usually measured on the horizontal axis and velocity on the vertical. These diagrams are useful for solving some motion problems and also provide a clear picture of what is happening. The diagrams do not show direction of the moving body, only its magnitude, and so we could really call them speed–time diagrams. See Figure 4.3.

A horizontal line on the diagram indicates constant velocity whereas a sloping line, as in Figure 4.4, indicates a change in velocity, i.e. an acceleration.

The upward slope from left to right in Figure 4.4 shows an increase in velocity, i.e. positive acceleration.

A downward slope from left to right (Figure 4.5) shows a decrease in velocity, i.e. negative acceleration. The area underneath the plotted line on the diagram represents the distance travelled. The formulae and theories we deal with assume that

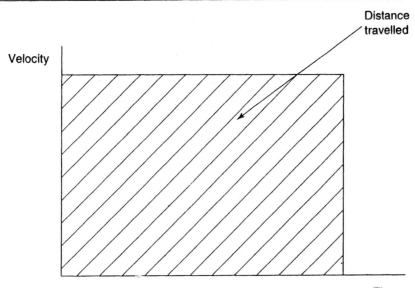

Figure 4.3

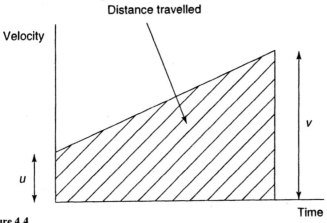

Figure 4.4

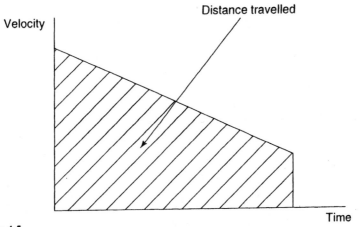

Figure 4.5

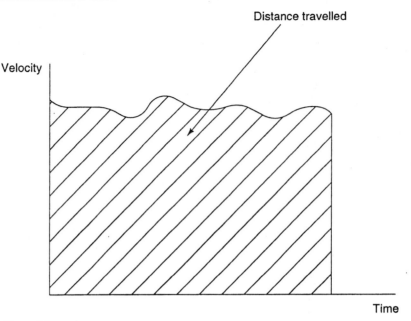

Figure 4.6

any acceleration is uniform and so there will be no curved lines on the graph as in Figure 4.6.

As usual the best way of demonstrating these diagrams is with an example.

Example 4.4

A van accelerates from rest to a velocity of 20 m/s in 20 s. It then travels at a constant velocity for 60 seconds before retarding to rest over a period of 40 s. Draw a velocity–time diagram. Using the diagram calculate the total distance travelled and the acceleration and retardation.

Mark a few values at even spaces on the vertical velocity axis, including 20 m/s. The time axis needs to go up to 120 s, the total time taken. Draw in the acceleration as a straight line from the origin (0,0) to co-ordinates of (20,20). Next draw in the constant velocity period as a straight horizontal line starting at the end of the acceleration line, making it 60 s long on the time axis. Finally, draw the retardation line as a straight line from the end of the constant velocity line down to 0 m/s over a length of 40 s on the time axis. You now have a clear picture of the motion of the van over its journey. This is shown in Figure 4.7.

As the area under the curve represents the distance travelled, split the diagram up into shapes whose areas are easy to calculate, such as two triangles and a rectangle. Now calculate the areas of the individual shapes and add them up.

$$Area_{accel} = 0.5 \times 20\,[m/s] \times 20\,[s] = \ \ 200\,m$$
$$Area_{const} = 20\,[m/s] \times 60\,[s] \qquad = 1200\,m$$
$$Area_{retard} = 0.5 \times 20\,[m/s] \times 40\,[s] = \ \ 400\,m$$
$$200\,m + 1200\,m + 400\,m = \textbf{1800 m}$$

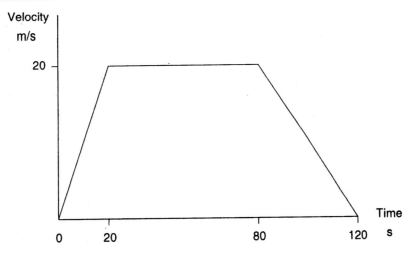

Figure 4.7

Acceleration is equal to the slope of the line.

$$\text{Acceleration} = \frac{\text{change in velocity}}{\text{time}}$$

$$= \frac{v - u}{t}$$

$$= \frac{20 - 0}{20}\left[\frac{\text{m/s}}{\text{s}}\right] = \frac{20\,[\text{m/s}]}{20\,[\text{s}]}$$

$$= 1\ \text{m/s}^2$$

$$\text{Retardation} = \frac{0 - 20}{40}\left[\frac{\text{m/s}}{\text{s}}\right]$$

$$= -0.5\ \text{m/s}^2$$

Use of vector diagrams and relative velocity

So far the velocities we have looked at have been the velocities of moving bodies relative to fixed points on the Earth. These are called **absolute velocities**. A **relative velocity** is the velocity of a body expressed relative to another body. You will probably be familiar with this; for example if you are in a car travelling at 70 mile/h in the middle lane of a motorway and another car goes past you in the outside lane at 100 mile/h, then to you the other car appears to pass you at 30 mile/h. The relative velocity of the other car to your car is 30 mile/h although the absolute velocity of the other car is 100 mile/h. The relative velocity of another passenger in your car to you is zero, although you are both travelling with an absolute velocity of 70 mile/h. These examples are simple enough as the absolute velocities all act in the same direction. When the velocities act in different directions vector diagrams are needed to solve them. A relative velocity is the *vector difference* of the two velocities.

$$v_{BA} = v_B - v_A$$

❑ *Example 4.5*

A bike travels at 20 m/s due north. A car travels at 30 m/s due west. Find the relative velocity of the bike to the car.

See Figures 4.8 and 4.9. The velocity of the bike relative to the car is the velocity of the bike minus the velocity of the car.

Velocity of the bike relative to the car = **36 m/s.**

The angle θ = 56° east of north.

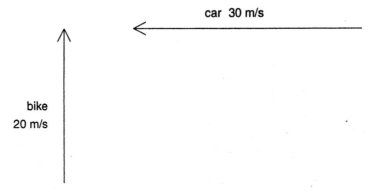

Figure 4.8

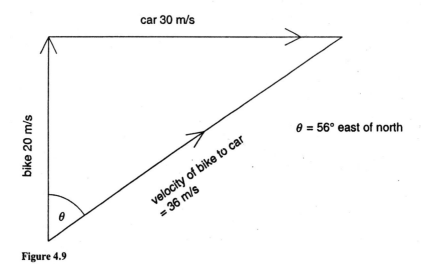

Figure 4.9

❑ *Example 4.6*

A car travels due north towards an overhead rail bridge with a velocity of 40 m/s. A train travels east over the bridge at 35 m/s. Find the velocity of the car relative to the train.

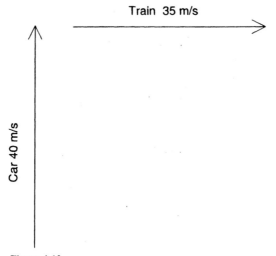

Figure 4.10

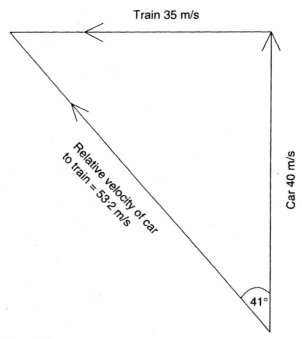

Figure 4.11

As we are finding the velocity of the car relative to the train then the vector diagram is drawn of the velocity of the car minus the velocity of the train. See Figures 4.10 and 4.11.

Velocity of car minus velocity of the train = **53.2 m/s, 41° west of north.**

4.2 Impulse, impact and momentum

When two bodies collide, the forces that act may only be effective for very brief periods but they can cause considerable momentum changes. These forces are called **impulse forces**. From Newton's second law:

Force = rate of change of momentum

We can write this as $F = \dfrac{m(v - u)}{t}$

$$\therefore F \times t = m(v - u)$$

The product $F \times t$ is the impulse of the force F, when F is applied for a short period of time t.

The expression $m(v - u)$ represents the change in momentum of mass m in time t.

So, impulse = change in linear momentum

The units of impulse are the same as for the change in momentum kg m/s, although sometimes the unit N s is used as 1 N = 1 kg m/s^2. If there is no impulse, then there is no change in linear momentum. This is expressed by another one of Newton's laws: the principle of conservation of momentum.

The perfectly elastic collision

To deal with the theory of the conservation of momentum we use the concept of an elastic body. An elastic body is one that can undergo impact with another elastic body with no loss of energy through heat or sound, etc. This is known as a **perfectly elastic collision**. With an elastic collision, the total energy before the collision is equal to the total kinetic energy after the collision. This can happen only if the kinetic energy is not converted into any other form of energy.

Principle of conservation of momentum

Momentum can be destroyed only by a force and can be created only by the action of a force. If no external force acts on a body or system of bodies, then the momentum remains constant (in both magnitude and direction).

If two elastic bodies travelling in the same direction collide they exert a force F on each other for a short period of time t. The force that the first body exerts on the second is equal and opposite to the force that the second body exerts on the first. Therefore an impulse, Ft, acts on each body causing a change in momentum in each. The increase in momentum of one body will equal the decrease in momentum of the other body. The sum of the momentum before the impact will equal the sum of the momentum after the impact. Except for perfectly elastic bodies there will always be some energy converted from kinetic to another form. For perfectly non-elastic bodies all energy is converted from kinetic into other forms, and the kinetic energy after the collision is zero. In reality the collision of the bodies lies somewhere between these two extremes.

Coefficient of restitution

The coefficient of restitution takes account of the level of elasticity of a body. It is represented by e and has a value between 0 and 1. For a perfectly elastic collision $e = 1$. For a perfectly non-elastic collision $e = 0$ and there is no rebound. It is used in calculation as follows:

relative velocity after impact = relative velocity before impact $\times -e$

$$v_1 - v_2 = (u_1 - u_2) \times -e$$

$$-e = \frac{v_1 - v_2}{u_1 - u_2}$$

If the masses and velocities of two moving bodies, A and B, are known then by the principle of conservation of momentum:

$$m_A u_A + m_B u_B = m_A v_A + m_B v_B$$

This gives one equation with two unknown values, v_A and v_B. The coefficient of restitution equation allows the unknown velocities to be determined.

☐ Example 4.7

A mass of 10 kg moves with a velocity of 20 m/s. This collides with a stationary mass of 15 kg. After the collision the 10 kg mass is stationary. Calculate:

1. the velocity of the 15 kg mass after the collision
2. the coefficient of restitution
3. the loss of kinetic energy due to impact.

$$m_A = 10 \text{ kg}; \quad u_A = 20 \text{ m/s}; \quad v_A = 0$$
$$m_B = 15 \text{ kg}; \quad u_B = 0 \text{ m/s}; \quad v_B = ?$$

momentum before impact = momentum after impact

$$m_A u_A + m_B u_B = m_A v_A + m_B v_B$$
$$(10 \times 0) + (15 \times v_B) = (10 \, [\text{kg}] \times 20 \, [\text{m/s}]) + (15 \, [\text{kg}] \times 0)$$
$$\Rightarrow 15 v_B = 200$$
$$\therefore v_B = \frac{200}{15} \left[\frac{\text{m}}{\text{s}}\right] = \textbf{13.333 m/s}$$

relative velocity after impact = relative velocity before impact $\times -e$

$$e = -\frac{v_A - v_B}{u_A - u_B}$$

$$1 = -\frac{(0 - 13.333)}{(20 - 0)} \left[\frac{\text{m/s}}{\text{m/s}}\right] = \textbf{0.668}$$

Kinetic energy before impact $= \frac{1}{2}(m_A u_A^2 + m_B u_B^2)$

$$= \frac{1}{2}\left(\left(10\,[\text{kg}] \times 20^2 \left[\frac{\text{m}^2}{\text{s}^2}\right]\right) + \left(15\,[\text{kg}] \times 0 \left[\frac{\text{m}^2}{\text{s}^2}\right]\right)\right)$$

$$= 2000 \left[\frac{\text{kg m}^2}{\text{s}^2}\right] = 2\,\text{kJ}$$

Kinetic energy after impact $= \frac{1}{2}(m_A v_A^2 + m_B v_B^2)$

$$= \frac{1}{2}\left(\left(10\,[\text{kg}] \times 0 \left[\frac{\text{m}^2}{\text{s}^2}\right]\right) + \left(15\,[\text{kg}] \times 13.333^2 \left[\frac{\text{m}^2}{\text{s}^2}\right]\right)\right)$$

$$= 1333.267 \left[\frac{\text{kg m}^2}{\text{s}^2}\right] = 1.333\,\text{kNm}$$

Loss of kinetic energy $= 2\,[\text{kJ}] - 1.3\,[\text{kJ}]$

$$= 0.667\,\text{kJ} = \mathbf{667\,J}$$

❏ *Example 4.8*

A mass of 5 kg moving at 10 m/s hits another mass of 8 kg with a velocity of 4 m/s moving in the same direction. The coefficient of restitution is 0.8. Calculate the velocities after the collision and the loss of kinetic energy.

$m_A = 5\,\text{kg}; \quad u_A = 10\,\text{m/s}; \quad v_A = ?$

$m_B = 8\,\text{kg}; \quad u_B = 4\,\text{m/s}; \quad v_B = ?$

relative velocity after impact = relative velocity before impact $\times -e$

$$e = -\frac{v_A - v_B}{u_A - u_B}$$

$$v_A - v_B = -e \times (u_A - u_B)$$

$$= -0.8 \times (10 - 4) \left[\frac{\text{m}}{\text{s}}\right]$$

∴ $\qquad v_A - v_B = -4.8\,\text{m/s}$ \hfill (1)

Remember that this is relative velocity, i.e. the velocity of mass A relative to mass B.

Momentum before impact = momentum after impact

$$m_A u_A + m_B u_B = m_A v_A + m_B v_B$$

$$5v_A + 8v_B = \left(5\,[\text{kg}] \times 10 \left[\frac{\text{m}}{\text{s}}\right]\right) + \left(8\,[\text{kg}] \times 4 \left[\frac{\text{m}}{\text{s}}\right]\right)$$

$$5v_A + 8v_B = 82\,\text{kg m/s} \hfill (2)$$

If we now multiply equation 1 by 5:

$$(v_A - v_B) \times 5 = -4.8 \times 5$$
$$5v_A - 5v_B = -24 \tag{3}$$

Now we can subtract equation 2 from 3 to get rid of the $5v_A$ term:

$$(5v_A - 5v_A) + (-5v_B - 8v_B) = -24 - 82$$
$$-13v_B = -106$$
$$\therefore v_B = \frac{-106}{-13} = \textbf{8.154 m/s}$$

Substitute this value back into equation 1:

$$v_A - v_B = -4.8 \left[\frac{m}{s}\right] \tag{1}$$

$$v_A - 8.154 = -4.8 \left[\frac{m}{s}\right]$$

$$\therefore v_A = -4.8 + 8.154 = \textbf{3.354 m/s}$$

Kinetic energy before impact $= \frac{1}{2}(m_A u_A^2 + m_B u_B^2)$

$$= \frac{1}{2}\left(\left(5\,[kg] \times 10^2 \left[\frac{m^2}{s^2}\right]\right) + \left(8\,[kg] \times 4^2 \left[\frac{m^2}{s^2}\right]\right)\right)$$

$$= 314.000 \left[\frac{kg\,m^2}{s^2}\right] = 314.000\,[J]$$

Kinetic energy after impact

$$= \frac{1}{2}(m_A v_A^2 + m_B v_B^2)$$

$$= \frac{1}{2}\left(\left(5\,[kg] \times 3.354^2 \left[\frac{m^2}{s^2}\right]\right) + \left(8\,[kg] \times 8.154^2 \left[\frac{m^2}{s^2}\right]\right)\right)$$

$$= 294.074 \left[\frac{kg\,m^2}{s^2}\right] = 294.074\,J$$

Loss of kinetic energy $= 314.000\,J - 294.074\,J$

$$= \textbf{19.926 J}$$

4.3 Rotational dynamics

So far we have been looking at linear motion, such as the motion of an engine piston. Another important type of motion is angular motion, such as the motion of a crankshaft. A set of formulae for angular motion can be derived in a similar manner to those for linear motion. In the chapter on 'Torsion', we looked at expressing an angle turned in radians. See Figure 4.12.

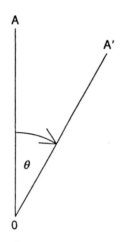

Figure 4.12

The crank OA turns in a clockwise direction about O. The displacement of the crank can be expressed as an angle θ measured in radians. The radian is an especially useful unit because it is then easy to convert angular motion to linear motion. If the crank turns 1 radian the end of the crank will move the distance of 1 radius.

Linear displacement, $S\,[\text{m}] = r\,[\text{m}] \times \theta^{[r]}$

Angular velocity is the rate of change of angular displacement and is represented by ω.

$$\text{Angular velocity, } \omega = \frac{\theta}{t}\left[\frac{\text{rad}}{\text{s}}\right]$$

Angular velocity should be expressed in radians per second for calculation purposes, but information in a problem is normally given in revolutions per minute. Angular acceleration is the rate of angular velocity and is represented by α.

$$\text{Angular acceleration } \alpha = \frac{\omega}{s}\left[\frac{\text{rad}}{\text{s}^2}\right]$$

As with linear acceleration we shall only be concerned with constant acceleration. Now take one of the linear equations of motion and apply it to the crank OA:

$$v = u + at \tag{2}$$

and substitute the linear motion variables for rotational ones. The linear velocity of A can be found from the angular velocity in radians per second multiplied by the radius of the crank OA:

$$v \left[\frac{m}{s} \right] = \omega_2 \left[\frac{rad}{s} \right] \times r \, [m]$$

In the same way:

$$u \left[\frac{m}{s} \right] = \omega_1 \left[\frac{rad}{s} \right] \times r \, [m]$$

$$\text{and } a \left[\frac{m}{s^2} \right] = \alpha \left[\frac{rad}{s^2} \right] \times r \, [m]$$

So $\omega_2 r = \omega_1 r + \alpha r t$

The radius can be cancelled as it is a multiple of each side:

$$\Rightarrow \omega_2 = \omega_1 + \alpha t$$

$$S = \frac{u + v}{2} \times t \tag{1}$$

$$\Rightarrow \theta r = \frac{\omega_1 r + \omega_2 r}{2} \times t$$

The radius, r, can be cancelled as it is a multiple of each side:

$$\Rightarrow \theta = \frac{\omega_1 + \omega_2}{2} \times t$$

$$S = ut + \tfrac{1}{2} a t^2 \tag{3}$$

$$\Rightarrow \theta r = \omega r t + \tfrac{1}{2} \alpha r t^2$$

Cancel the radius, r:

$$\Rightarrow \theta = \omega_1 t + \tfrac{1}{2} \alpha t^2$$

$$v^2 = u^2 + 2aS \tag{4}$$

$$\Rightarrow (\omega_2 r)^2 = (\omega_1 r)^2 + 2 \alpha r \theta r$$

$$\Rightarrow \omega_2^2 = \omega_1^2 + 2 \alpha \theta$$

We now have a set of equations for dealing with rotational dynamics.

☐ *Example 4.9*

A car accelerates from 9 m/s (about 20 mile/h) to 36 m/s (about 80 mile/h) in 6 s. The wheel radius is 0.35 m. Calculate the linear acceleration of the car and calculate the rotational acceleration of the wheels.

From $v = u + at$

$$\Rightarrow a = \frac{v - u}{t}$$

$$= \frac{36\,[\text{m/s}] - 9\,[\text{m/s}]}{6\,[\text{s}]} = 4.5\;\text{m/s}^2$$

The speed of the road wheels can be calculated by dividing the velocity of the car by the circumference of the wheels (circumference = πd).

$$n_1 = \frac{9\,[\text{m/s}]}{2 \times 0.35\,[\text{m}] \times \pi} = 4.093\;\text{rev/s}$$

$$4.093\left[\frac{\text{rev}}{\text{s}}\right] \times 2\pi\left[\frac{\text{rad}}{\text{rev}}\right] = 25.717\;\text{rad/s}$$

$$n_2 = \frac{36\,[\text{m/s}]}{0.7\,[\text{m}] \times \pi} = 16.370\;\text{rev/s}$$

$$16.370\left[\frac{\text{rev}}{\text{s}}\right] \times 2\pi\left[\frac{\text{rad}}{\text{rev}}\right] = 102.856\;\text{rad/s}$$

The calculations have been done like this to make the method clear but the quickest way would have been to divide the vehicle speed by the wheel radius to obtain the angular velocity directly in radians,

$$\text{e.g.} \quad n_1 = \frac{9\,[\text{m/s}]}{0.35\,[\text{m}]} = 25.717\;\text{rad/s}$$

from $\omega_2 = \omega_1 + \alpha t$

we get $\alpha = \dfrac{\omega_2 - \omega_1}{t}$

$$= \frac{102.856 - 25.717}{6}\left[\frac{\text{rad}}{\text{s}}\right]\left[\frac{1}{\text{s}}\right] = 12.857\;\text{rad/s}^2$$

When a body undergoes a linear acceleration the external force F is applied to the centre of gravity G, as shown in Figure 4.13.

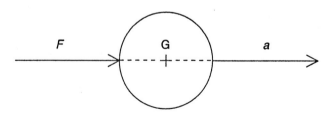

Figure 4.13

As the force F has no turning moment about the centre of gravity G then the body does not rotate. Every particle of the body moves with the same linear acceleration. Consider a force F applied to a small mass m on the end of a very light arm r that can

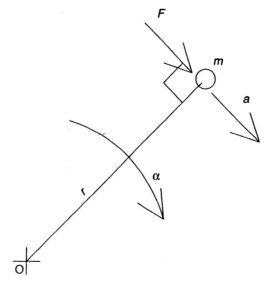

Figure 4.14

pivot about O. The force acts perpendicular to the arm r. Therefore a torque is applied to the arm r about the axis, O (see Figure 4.14).

The mass at this instant is given a linear acceleration a such that $F = m \times a$. The linear acceleration a is equal to the angular acceleration α multiplied by the radius r.

$$a = \alpha \times r$$

$$\therefore F = m \times \alpha \times r$$

The moment of the force is the torque T applied to the arm:

$$T = F \times r$$

$$\therefore T = (m \times \alpha \times r) \times r$$

$$= m \times \alpha \times r^2$$

$$= mr^2 \times \alpha$$

The term mr^2 is an important quantity in rotational dynamics. It is the second moment of mass about O, commonly called the **moment of inertia** and given the symbol I. The units are kg m^2.

The moment of inertia I of a whole body rotating is the sum of the second moment of mass values of all the individual particles of mass:

$$T = I\alpha$$

The moment of inertia is a measure of a body's resistance to rotational acceleration and depends upon the mass and the distribution of the mass. It is comparable to mass in linear motion. The formula:

$$T = I\alpha$$

for rotational dynamics is comparable to:

$$F = ma$$

If the mass of a rotating body could be assumed to be concentrated at a particular radius k from the axis, then $I = mk^2$. The radius k of a rotating body is called the **radius of gyration**. k is used rather than r to avoid confusion with other radius values of the body. For calculation purposes a rotating body of any shape could be replaced by a thin ring of radius k.

Look at Figure 4.15 and at Figure 4.16. In Figure 4.16, the total mass is still m but it is all concentrated at the radius k. Two different rotating bodies can have identical masses but completely different values for radius of gyration. A flywheel is designed so that the mass is distributed as far away as possible to give a higher moment of inertia I.

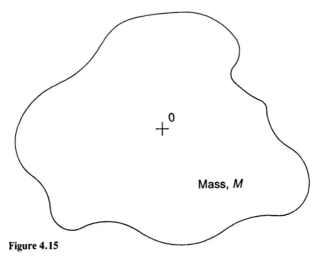

Figure 4.15

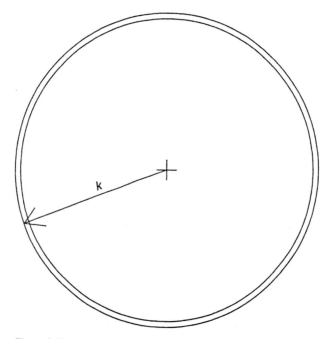

Figure 4.16

☐ *Example 4.10*

A flywheel has a mass of 18 kg and a radius of gyration of 0.2 m. Calculate the torque required to give an acceleration of 10 rad/s².

$$T = I\alpha$$

$$I = mk^2$$

$$= 18\,[\text{kg}] \times 0.2^2\,[\text{m}^2] = 0.72 \text{ kg m}^2$$

$$T = 18\,[\text{kg}] \times 0.2^2\,[\text{m}^2] \times 10\left[\frac{\text{rad}}{\text{s}^2}\right] = 7.2\left[\frac{\text{kg m}^2}{\text{s}^2}\right]$$

$$= \textbf{7.2 Nm}$$

Rotational momentum

Consider a small rotating mass, m. For linear motion:

force = rate of change of linear momentum

and force × time = change of linear momentum

= impulse

The equivalent formula for angular motion is:

torque = rate of change of angular momentum

torque × time = change of angular momentum

As torque = force × distance perpendicular from the force to axis

= moment of force about the axis

then **angular momentum** = moment of linear momentum about axis

The linear momentum of a rotating mass = mass × linear velocity

= $m \times v$

In rotational terms this is $m \times \omega r$

When considering a whole body the sum of mr^2 for all mass particles is the moment of inertia I.

So, moment of momentum = moment of inertia × angular velocity

or, angular momentum = $I\omega$

The units are: $[\text{kg m}^2] \times \left[\dfrac{\text{rad}}{\text{s}}\right] = \left[\dfrac{\text{kg m}^2}{\text{s}}\right]$

Note that the units of angular momentum are different from the units of linear momentum and so the two cannot be added together in a system to obtain total momentum. For a rotating body with a moment of inertia I whose angular velocity changes from ω_1 to ω_2,

change of angular momentum = $I(\omega_2 - \omega_1)$

For a change in angular momentum occurring in time t:

$T \times t$ = change of angular momentum

= angular impulse of the torque, T.

This gives the relationship between angular impulse and angular momentum. The angular impulse of a torque acting on a body for a given time t equals the change in angular momentum of the body in this time. The units of angular impulse are the same as angular momentum, kg m²/s, or, from torque × time, Nms.

You can check this as follows:

$$[\text{Nms}] = \left[\frac{\text{kg m}^2}{\text{s}}\right]$$

$$\text{As } 1\,[\text{N}] = 1\left[\frac{\text{kg m}}{\text{s}^2}\right]$$

$$\text{then } [\text{Nms}] = \left[\frac{\text{kg m}}{\text{s}^2}\right] \times [\text{m}] \times [\text{s}] = \left[\frac{\text{kg m}^2}{\text{s}}\right]$$

The principle of conservation of momentum can also be applied to angular systems. If the externally applied torque to any system is zero then the angular momentum remains constant.

❑ *Example 4.11*

A crankshaft and a flywheel have a total moment of inertia of 4 kg m² and rotate at 1200 rev/min. Calculate the angular momentum. The clutch *suddenly* engages the flywheel to a stationary driveshaft. The driveshaft system has a moment of inertia of 5 kg m². Calculate the common velocity of the new system and the impulse of each shaft.

$$\text{Velocity of crank}, \omega_c = 1200\left[\frac{\text{rev}}{\text{min}}\right] \times 2\pi\left[\frac{\text{rad}}{\text{rev}}\right] \times \frac{1}{60}\left[\frac{\text{min}}{\text{s}}\right]$$

$$= 125.664\left[\frac{\text{rad}}{\text{s}}\right]$$

$$\text{Angular momentum of crank} = I_c\omega_c$$

$$= 125.664\left[\frac{\text{rad}}{\text{s}}\right] \times 4\,[\text{kg m}^2]$$

$$= \textbf{502.655 kg m}^2\textbf{/s}$$

There is no externally applied torque on the system so the angular momentum before engagement is equal to angular momentum after engagement. After engagement the total system moves with a common angular velocity, ω. The initial velocity of the driveshaft system is zero.

So, $I_c\omega_c + I_d\omega_d = (I_c + I_d) \times \omega$

As $I_d\omega_d = 0$

$$502.655\left[\frac{\text{kg m}^2}{\text{s}}\right] = (4 + 5)\,[\text{kg m}^2] \times \omega$$

$$\therefore \omega = 502.655 \left[\frac{\cancel{kg} \, \cancel{m^2}}{s} \right] \times \frac{1}{(4+5) \, [\cancel{kg} \, \cancel{m^2}]}$$

$$= \mathbf{55.850 \ rad/s}$$

Angular impulse of crankshaft and flywheel:

angular impulse $(T \times t)$ = change in angular momentum

$$= I_c(\omega_2 - \omega_2)$$

$$= 4 \, [\text{kg m}^2] \times (55.850 - 125.664) \left[\frac{\text{rad}}{s} \right]$$

$$= .-\mathbf{279.25 \ kg \ m^2/s}$$

Angular impulse of the drive system:

angular impulse $(T \times t)$ = change in angular momentum

$$= I_d(\omega_2 - \omega_2)$$

$$= 5 \, [\text{kg m}^2] \times (55.850 - 0) \left[\frac{\text{rad}}{s} \right]$$

$$= +\mathbf{279.25 \ kg \ m^2/s}$$

Hence the angular momentum received by each shaft is equal and opposite to the other (Newton's third law). The crankshaft decelerates as the driveshaft accelerates.

4.4 Friction

Introduction

When any two surfaces are in contact a force exists that opposes relative motion between them. This opposition force is called **friction**. This is true even if one of the surfaces happens to be a fluid – a major part of the force required to move a ship through water is due to friction of the wetted surface.

For vehicles, problems involving friction fall into two groups: first where friction must be kept to a minimum, and secondly, where the maximum amount of friction is wanted.

The first group refers to machines such as engines or gearboxes where frictional forces must be kept to a minimum. Reducing friction reduces power loss and reduces wear on the bearing surfaces. This is usually achieved by using a fluid lubricant such as oil. The surfaces in relative movement are kept apart and there is little or no metal to metal contact. The frictional resistance then becomes a problem of fluid friction. Engine bearings are lubricted by oil flowing continuously through a system. The oil flow is caused by a pump and carries away heat generated. If the oil pump fails, a running engine usually seizes rapidly. The frictional resistance increases rapidly without the oil film and high temperatures develop, resulting in white metal melting and component surfaces welding. Engine treatments are now available in the form of a fluid that is added to the oil system after an oil change. The fluid deposits a permanent thin film of plastic called polytetrafluoroethylene (PTFE) on the metal surfaces in the oil system. When the oil is changed the plastic deposit remains on the surfaces. This has a very low coefficient of friction, reducing frictional forces and component wear. It also provides some protection to the engine in the event of an oil pump failure.

For the second class of problems, friction must be kept to a maximum. Lubrication must be avoided, and the contact surfaces must be kept clean and dry. These are the types of problems we will be looking at, and they are very important and include such things as brakes, tyre grip on the road, driving belts on pulley wheels, and clutches. Consider the disastrous effects of lubrication reducing friction in these situations – oil on your brake pads or riding a bike through a diesel spillage on the road. The study of friction here will be restricted to dry surfaces.

Coulomb friction

There are several different types of friction but the main one of interest to us is called coulomb friction. This is sometimes called dry friction, because it applies to clean dry surfaces. One way of considering friction is shown in Figure 4.17.

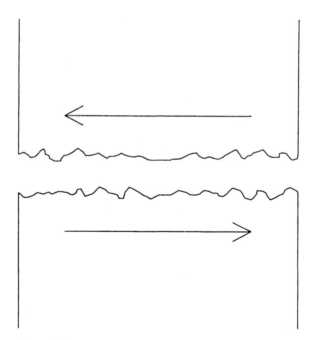

Figure 4.17

Surfaces all have tiny irregularities, (no matter how smooth the surface). When two surfaces move across one another, these irregularities interlock and cause a force that resists the motion of the surfaces. The force always acts parallel to the contact surfaces and in a direction that opposes the resultant motion. The force is called the force of friction and is usually represented by F_f.

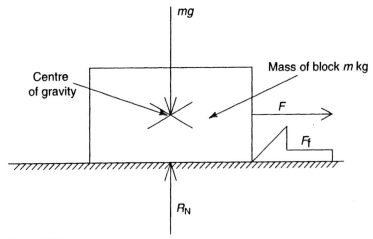

Figure 4.18

Think of a rectangular block of mass m kg resting on the horizontal ground (see Figure 4.18). The block will not move unless a force is applied to it. As the block is not moving then the forces present must be in equilibrium. There are two initial forces acting on the block. These are the force of gravity mg acting vertically downwards throughout the centre of gravity and the normal reaction R_N of the ground acting vertically up. In Chapter 2, the idea of a normal reaction was introduced. This is the upward reaction force that the ground applies to the block. This balances the gravitational force. If we now pull the block a little in a horizontal direction (force F) but not enough to actually move it, a fourth resisting force F_f is produced due to the action of the block and the ground trying to rub together – a frictional force. Remember that the force of friction always acts in a direction to prevent or resist the motion of the body. If we gradually pull the block a bit harder, increasing F, then the resisting friction force F_f will increase so that F and F_f will remain numerically equal. This process will continue until a maximum value of F_f is reached, beyond which it cannot increase. A further increase in F will cause the block to move in the direction in which F acts.

Admittedly, you will probably not come across a problem involving pushing a rectangular block across a horizontal surface but this is a useful example to help simplify the concept and thus make it easier to understand. The frictional force that prevents the block from initially moving is called 'static friction' (some people call this 'stiction'). The resisting frictional force that occurs when the block is moving is called the frictional force of motion or kinetic force of friction. When these two frictional forces are being represented, static friction is usually written as $F_{f(s)}$ and the frictional force of motion as $F_{f(k)}$. Both static and kinetic friction forces are coulomb or dry friction. Look at Figure 4.19. This is a plot of the dry or coulomb frictional force between the block and the ground against the pulling force F.

The highest value of static friction reached just before the block starts to move is called the limiting friction force. Kinetic friction is slightly lower than this value. This is why, when a car is made to do an emergency stop, the greatest braking effect is achieved by braking as hard as possible without letting the tyres skid on the road. Some braking systems are now designed to exploit this effect and prevent the wheels from locking completely when the brake pedal is pressed hard.

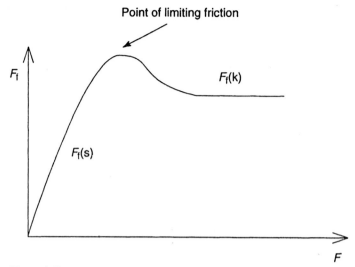

Point of limiting friction

Figure 4.19

Laws of friction

There are some laws of friction you need to be aware of. These were first investigated by a French scientist in the 18th century called Coulomb, which is where the name comes from for this type of friction. The laws are only approximate but accurate enough for most problems. Do not try to learn them off by heart at this stage. Once you have dealt with a few examples they will become familiar.

1. The frictional force always opposes motion and acts parallel to the surfaces. You saw this in the sliding block example.
2. The magnitude of the frictional force F_f is directly proportional to the perpendicular force between the two surfaces. With the sliding block this was the force of gravity mg or the normal reaction R_N.
3. The frictional force F_f is independent of the area of contact. This means that, if we turned the block onto one of its smaller ends and pushed in the same way, the resisting force F_f would be the same, provided that the surface of the end of the block was the same as the original side.
4. The frictional force F_f is independent of the speed of motion. Notice in Figure 4.19 that the line is horizontal for $F_{f(k)}$, provided that the velocity is not too high.
5. The limiting friction of $F_{f(s)}$ is greater than the kinetic friction $F_{f(k)}$ – which you already knew.
6. The frictional force F_f depends on the materials in contact and the condition of their surfaces.

There are a few exceptions to these rules under extreme circumstances. For example, if the force between the two surfaces (e.g. mg) becomes too high then the surfaces can become welded together. These exceptions need not worry us for most problems.

The coefficient of friction

There are two coefficients of friction. These are very useful to us. They are the coefficient of static fricion, μ_s, and the coefficient of kinetic friction μ_k. These are based on law 2 but only apply once the point of limiting friction has been reached. The coefficient of static friction is the ratio between the maximum frictional force F_f and the normal reaction R_N.

Coefficient of static friction, $\mu_s = \dfrac{F_{f(k)}}{R_N}$

Once sliding is taking place there is a constant ratio between the frictional force F_f and the normal reaction R_N. This is known as the coefficient of kinetic friction, μ_k.

Coefficient of kinetic friction, $\mu_k = \dfrac{F_{f(s)}}{R_N}$

☐ *Example 4.12*

A tool box with a flat base which weighs 600 N is dragged across a concrete floor. A horizontal force of 150 N must be applied to it before it starts to move. After it has started to move a horizontal force of 120 N is sufficient to keep it moving at a steady speed. Calculate the coefficients of static and kinetic friction.

When the tool box is on the point of moving, the applied force must equal the force of limiting friction, i.e. $F_{f(s)} = 150$ N.
   The normal reaction force exerted by the floor on the tool box is equal to the weight of the box, i.e.

$$R_N = 600 \text{ N}$$

Therefore, $\mu_s = \dfrac{F_{f(s)}}{R_N} = \dfrac{150\ [\cancel{N}]}{600\ [\cancel{N}]} = \mathbf{0.25}$

Notice that the coefficient has no units as the newtons cancel in the equation.

$$\mu_k = \frac{F_{f(k)}}{R_N} = \frac{120\ [\cancel{N}]}{600\ [\cancel{N}]} = \mathbf{0.2}$$

If it takes a maximum pull of 4 N to open an empty drawer in the tool box that weighs 15 N, then how much of a pull is needed to open the drawer after you have put spanners in it weighing 40 N?

$$\mu_s = \frac{F_{f(s)}}{R_N} = \frac{4\ [N]}{15\ [N]} = 0.267$$

The new total weight of the drawer $= 15\ [N] + 40\ [N] = 55$ N

$$F_{f(s)} = \mu_s \times R_N = 0.267 \times 55\ [N] = 14.7 \text{ N}$$

So the new force needed to open the drawer is **14.7 N**.

Angle of friction

Go back to the rectangular block on the ground being pushed. You can consider that two of the forces, F_f and R_N, are exerted by the ground on the block. The resultant of these two forces R is the total reaction exerted by the ground on the block.

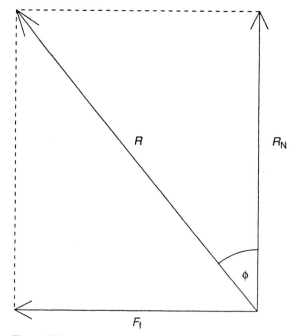

Figure 4.20

In Figure 4.20 the angle ϕ (pronounced phi) that the resultant reaction makes with the normal reaction is known as the angle of friction. As F_f and R_N are at right angles then:

$$\tan \phi = \frac{F_f}{R_N}$$

You may have noticed that this is the same formula as for μ_s and μ_k, so we can say that:

at the point of limiting friction $\quad \mu_s = \tan \phi$
and when sliding takes place $\quad \mu_k = \tan \phi$

Now the four forces present with the sliding block can be replaced by three forces, R, F and mg, and a force triangle drawn (Figure 4.21). The tangent of the angle of friction is equal to the coefficient of friction for the surfaces. Notice that the resultant of F_f and R_N, which is R, is inclined in the opposite direction to that in which motion occurs or is about to occur.

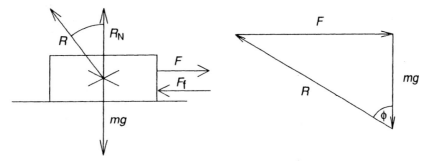

Figure 4.21

❑ *Example 4.13*

A force of 500 N is required to pull an engine of mass 200 kg across a floor at a steady speed. The force F acts upwards at an angle of 20° to the horizontal. What is the coefficient of friction between the engine and the floor?

See Figure 4.22. In order to simplify the problem the force applied F can be resolved into its vertical and horizontal components. The vertical component is equal to $F \sin 20°$ (see page 37) and the horizontal component is equal to $F \cos 20°$. The vertical component of the force will reduce the force between the two sliding surfaces.

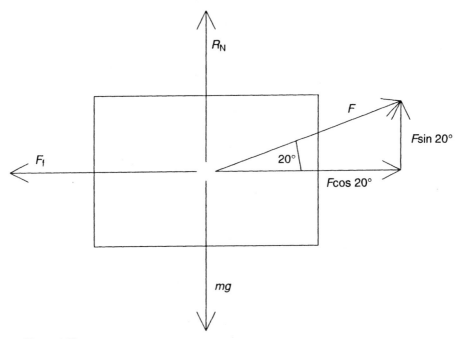

Figure 4.22

Force between surfaces = weight − vertical component

$$R_N = m\,g - F \sin 20°$$

$$= \left(200\,[\text{kg}] \times 9.81\left[\frac{\text{m}}{\text{s}^2}\right]\right) - (500\,[\text{N}] \times \sin 20°)$$

$$= 1962.000\,[\text{N}] - 171.010\,[\text{N}]$$

$$= 1790.990\,\text{N}$$

The force necessary to overcome the frictional resistance is equal to the horizontal component of the applied force.

$$F_f = F \cos 20°$$

$$= 500\,[\text{N}] \times \cos 20°$$

$$= 469.846\,\text{N}$$

As the block is sliding then we are talking about the coefficient of kinetic friction.

Coefficient of kinetic friction, $\mu_k = \dfrac{F_f}{R_N}$

$$= \frac{469.846}{1790.990} = 0.262$$

The inclined plane and the screw thread

Imagine a block of mass m resting on a slope. There is negligible friction between the bottom surface of the block and the slope.

See Figure 4.23. The weight W of the block acts vertically down and is equal to mg.

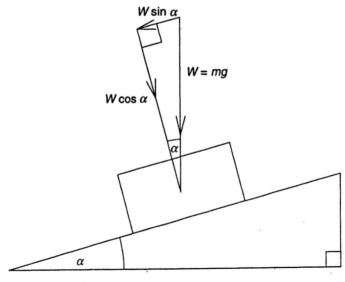

Figure 4.23

The angle of the slope to the horizontal is represented by the angle α. The block would tend to slide down the slope due to a component of the vertical force, W:

component of W acting down the slope = $W \sin \alpha$

The normal reaction between the block and the slope is equal to the component of the force W that acts at right angles to the slope:

component of W acting normally to the slope = $W \cos \alpha$

The weight W has two effects, one that tends to move the block down the slope and another that creates a normal reaction between the block and the slope.

Now, if we take friction into account and pull the block up the slope by a force parallel to the slope, two forces must be overcome: a force due to gravity and a force due to frictional resistance. See Figure 4.24.

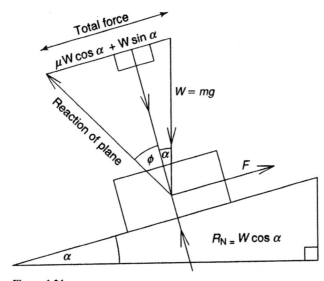

Figure 4.24

If $\mu = \dfrac{F_f}{R_N}$ then to calculate the frictional force resisting motion,

$$F_f = \mu R_N$$

Remember that the frictional force F_f opposes motion and will therefore act parallel to the slope. We know that the normal reaction between the block and the slope, R_N, is equal to $W \cos \alpha$. Therefore the frictional force F_f is equal to:

$$F_f = \mu W \cos \alpha$$

The reaction of the slope on the block acts at an angle of ϕ, which is the angle of friction, to the normal reaction between the two surfaces. The total force to pull the block,

F = force to overcome gravity + frictional resistance force

$$F = W \sin \alpha + \mu W \cos \alpha$$

❏ *Example 4.14*

A block of weight 50 N is to be pulled up a slope inclined at 30°. The coefficient of friction between the slope and the block is 0.2. Find the force required to pull the block up the slope if it acts parallel to the slope.

$$\text{The total force, } F = W \sin \alpha + \mu W \cos \alpha$$
$$= (50\,[N] \times \sin 30°) + (0.2 \times 50\,[N] \times \cos 30°)$$
$$= 25.000\,[N] + 8.660\,[N]$$
$$= \mathbf{33.660\,[N]}$$

When a nut is turned on a bolt, the problem is similar to a block climbing up a slope. We will consider a square section screw thread as it makes the example simpler. A nut is used on the bolt to push against an axial load, W. If the bolt is positioned vertically the force to turn the nut is assumed to be a horizontal one. The load W then acts vertically.

Imagine one of the screw threads (shown in Figure 4.25) unwound.

Figure 4.25 Square screw thread

The bearing surface of the nut and bolt is like an inclined surface wrapped around the shaft (Figure 4.26). The length of the slope is equal to the length of the helix of the thread. The horizontal component of this slope is equal to the mean circumference of the thread. The vertical component is equal to the pitch of the thread. To understand how this works, we first need to understand the vertical and horizontal forces involved when a block is pushed up an incline with a horizontal force.

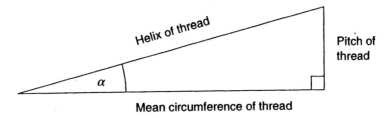

Helix of thread

Pitch of thread

Mean circumference of thread

Figure 4.26

Look at Figure 4.27. The total vertical component is the weight W. The horizontal component X is the *force* that acts at right angles to the weight. The total angle between the weight and the reaction of the slope is equal to $(\phi + \alpha)$. The horizontal component X necessary to pull the block up the slope can be found from:

$$\tan(\phi + \alpha) = \frac{\text{opposite side}}{\text{adjacent side}} = \frac{\text{horizontal component, } X}{\text{Weight, } W}$$

Therefore, horizontal component, $X = \tan(\phi + \alpha) \times W$

If you carry out the following calculations but for the horizontal force pulling the block down the slope you will find that:

horizontal component, $X = W \tan(\phi - \alpha)$

This is because the frictional resistance acts in the opposite direction.

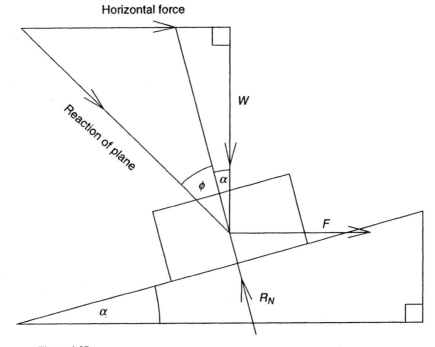

Horizontal force

Reaction of plane

W

ϕ α

F

R_N

α

Figure 4.27

Now think back to the unwound screw thread. If the average diameter of the screw is d then the length of the unwound circumference is πd. The pitch P is the length moved along the screw if you follow exactly one turn. The torque required to turn something is the force required, multiplied by the radius at which it is applied. For a torque T applied to a screw thread, the force is always applied perpendicular to the screw, which is the reason why the sliding block was considered with a horizontal force pushing it. We can now apply the formula for that force to the screw thread.

$$T_{screw} = X \times \text{mean radius}$$

$$= W \tan(\phi \pm \alpha) \times \frac{d}{2}$$

ϕ in this case is still the angle of friction, which is $\tan^{-1}\mu$ and α is the angle of the slope which is $\tan^{-1}(P/\pi d)$. The coefficient of friction between the nut and the screw will affect the angle of friction and consequently the torque required to turn the nut. You will be familiar with how much easier it is to undo a clean lightly greased nut and bolt compared with the same type of nut and bolt that is corroded and dirty.

❏ Example 4.15

A screw jack with a square thread of mean diameter 50 mm needs to lift a load of 1000 N. The thread is single start and has a pitch of 10 mm. The coefficient of friction for the nut and the screw is 0.2. Calculate the torque required to turn the nut.

$$\phi = \tan^{-1}\mu = \tan^{-1}0.2 = 11.31$$

$$\alpha = \tan^{-1}(P/\pi d) = \tan^{-1}(10/\pi 50) = 3.64$$

$$T_{screw} = W \tan(\phi \pm \alpha) \times \frac{d}{2}$$

$$= 1000\,[\text{N}] \times \tan(11.31 + 3.64) + \frac{50 \times 10^{-3}}{2}\,[\text{m}]$$

$$= 6.68\,\text{Nm}$$

The friction clutch

The internal combustion engine produces significant power only at high speeds. The engine must revolve at a speed that will produce sufficient power before connection of the drive to the wheels is made. The clutch allows the drive torque to be taken up gradually without reducing the engine speed too much, when movement commences from a stationary position. The drive needs to be disengaged periodically to change gear. A friction clutch is used in motor vehicles for transmitting torque from the engine to the gearbox, allowing for disengagement during gear change or when starting. The friction clutch uses the friction force that is produced when the driving and driven elements are pressed together by springs to produce torque. When the driving and driven element are held apart no drive is made. A common arrangement of a plate clutch is shown in Figure 4.28.

The engine drives the flywheel. The torque of the flywheel needs to be transmitted to the driven shaft. The driven plate is free to slide along the splined driven shaft. A

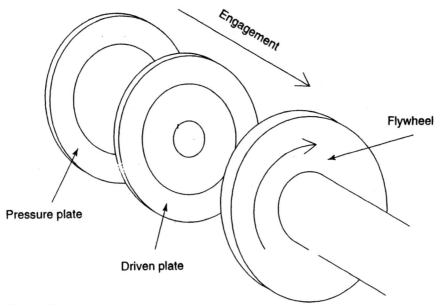

Flywheel

Pressure plate

Driven plate

Figure 4.28

special friction material is riveted to either side of this plate. When the clutch is engaged, a diaphragm spring in the clutch cover forces the pressure plate onto the friction material of the driven plate. The driven plate moves along the driven shaft a short distance and also comes into contact with the flywheel. The driven plate is then sandwiched between the flywheel and the pressure plate, and the torque is transmitted. When the clutch is engaged the flywheel and the driven plate must rotate at the same speed without slipping. So there needs to be sufficient force and friction to cause this. As the clutch pedal is pressed to disengage the clutch, a withdrawal sleeve allows the clutch forks to push the pressure plate away from the flywheel. Torque can then no longer be transmitted from the flywheel to the driven shaft. The clutch is engaged and disengaged gradually by controlling the force that the spring applies, pushing the plates together. There is a maximum torque that the clutch can transmit. The torque produced by the friction must be greater than the resisting torque of the driven shaft. When the maximum torque is exceeded the driven plate slips between the flywheel and pressure plate. This provides some protection against damage of the components and allows the drive to be engaged smoothly. This clutch has only one plate but some can have several.

The torque and power transmitted by a clutch can be calculated as follows:

axial thrust = pressure on friction material × friction material surface area

$$= p \left[\frac{N}{m^2} \right] \times A \ [m^2]$$

The friction material has a maximum pressure that it can be subjected to without damage. This obviously affects the design of the clutch. For example, to enable a certain axial thrust to be applied using a material with a low maximum pressure, a larger contact area of the material must be used than for a material with a higher maximum allowable pressure.

The axial thrust acts normally to the friction material surface and so can be represented by R_N.

Using $F_f = \mu R_N$, the frictional driving force can be calculated from:

$$F_f\,[\text{N}] = \mu \times p \times A\,[\text{N}]$$

The torque transmitted equals the frictional force multiplied by the mean radius of the friction material surface (see page 42).

$$\text{Torque} = F_f \times r$$
$$= \mu \times p \times A \times r\,[\text{Nm}]$$

The power transmitted is equal to the torque transmitted, multiplied by the shaft speed in radians per second.

$$\text{Power}\,[\text{W}] = T\,[\text{Nm}] \times \omega \left[\frac{\text{rad}}{\text{s}}\right]$$

❑ *Example 4.16*

A single plate clutch with a mean radius of 0.11 m has a friction material with a coefficient of friction of 0.3 and a contact area of 23.5×10^{-3} m². The pressure on the friction material is 130 kN/m². Calculate

(a) the frictional driving force acting at the mean radius
(b) the torque transmitted
(c) the power transmitted when the shaft speed is 2500 rev/min.

The frictional driving force,

$$F_f\,[\text{N}] = \mu \times p \times A\,[\text{N}]$$
$$= 0.3 \times 130 \left[\frac{\text{kN}}{\text{m}^2}\right] \times 10^3 \left[\frac{\text{N}}{\text{kN}}\right] \times 23.5 \times 10^{-3}\,[\text{m}^2]$$
$$= \mathbf{916.5\,N}$$

$$\text{Torque} = F_f \times r$$
$$= \mu \times p \times A \times r\,[\text{Nm}]$$
$$= 916.5\,[\text{N}] \times 0.11\,[\text{m}]$$
$$= \mathbf{100.815\,[Nm]}$$

$$\text{Power}\,[\text{W}] = T\,[\text{Nm}] \times \omega \left[\frac{\text{rad}}{\text{s}}\right]$$

Here we need the shaft speed in radians per second and it is given in revolutions per minute. Prepare a unity bracket before starting the calculation. Radians are explained fully on page 104. For the moment though one radian is an angle of about 57°.

In one revolution, i.e. 360°, there are 2π radians.

$$1\,\text{rev} = 2\pi\,\text{rad}$$

Divide both sides by seconds:

$$1\left[\frac{\text{rev}}{\text{s}}\right] = 2\pi\left[\frac{\text{rad}}{\text{s}}\right]$$

The revolutions per second need to be converted to revolutions per minute:

$$1\left[\frac{\text{rev}}{\text{s}}\right] \times 60\left[\frac{\text{s}}{\text{min}}\right] = 2\pi\left[\frac{\text{rad}}{\text{s}}\right]$$

$$1 = \frac{2\pi}{60}\left[\frac{\text{rad/s}}{\text{rev/min}}\right] = \frac{60}{2\pi}\left[\frac{\text{rev/min}}{\text{rad/s}}\right]$$

These are the two forms of the unity bracket for converting rev/min to rad/s or vice versa. When you are using a unity bracket, the unit that you want to get rid of is always on the bottom and the unit you want to keep on the top. We want the shaft speed in rad/s but are given it in rev/min so we use $2\pi/60[(\text{rad/s})/(\text{rev/min})]$.

$$\text{Power [W]} = 100.815\,[\text{Nm}] \times 2500\left[\frac{\text{rev}}{\cancel{\text{min}}}\right] \times \frac{2\pi}{60}\left[\frac{\text{rad/s}}{\text{rev/}\cancel{\text{min}}}\right]$$

$$= 26.393 \times 10^3\,\text{Nm/s} = 26.393\,\text{kW}$$

(1 W = 1 Nm/s = 1 J/s; see page 183)

Disc brakes

A similar approach can be used to solve problems on disc brakes. A number of pads are pressed onto the rotating cast iron disc to reduce the disc's angular velocity. The pad assembly is attached to the axle casing or suspension. The brake operating system is usually hydraulic. When the footbrake is pressed, a master cylinder pressurises a hydraulic system which transmits the pressure to wheel cylinders acting on the pads. The force acting on each pad depends upon the area of the piston in the wheel cylinder: the larger the area the larger the force. When the foot brake is released, return springs cause the fluid to flow back to the master cylinder. Two opposed pistons can be used, acting directly on two pads on each side of the disc; or a single piston can be used, with two pads sandwiched between the action of the piston and the reaction of the piston housing.

Even four pistons can be used when a greater degree of safety is required.

Force acting per brake pad,

F = hydraulic oil pressure × cross-sectional area of each piston

$F = p \times A$

Frictional force per pad $= \mu \times p \times A$ [N]

Total braking force per disc

$$F_f\,[\text{N}] = n \times \mu \times p \times A\,[\text{N}]$$

The friction torque acting on each brake disc is equal to the total braking force per disc multiplied by the mean radius of the pads:

$$T\,[\text{Nm}] = n \times \mu \times p \times A \times r\,[\text{Nm}]$$

The kinetic energy is dissipated as heat at the brake pads.
The work done by the disc is equal to the torque multiplied by the angle turned through in radians:

$$\text{work done} = T\,[\text{Nm}] \times \theta^{[r]}$$

The power transmitted (work done per second) is equal to the torque multiplied by the disc speed in radians per second:

$$\text{power} = T\,[\text{Nm}] \times \omega\,[\text{rad/s}]$$

This is equal to the heat generated per second.

☐ Example 4.17

A disc brake rotates at a velocity of 250 rev/min. A pair of friction pads act at a mean radius of 0.15 m and the coefficient of friction between the pads and the disc is 0.32. The diameter of the brake cylinder is 50 mm ($= 0.05$ m) and the hydraulic oil pressure is 600 kN/m². Calculate:
(a) the total braking force on the disc
(b) the torque acting on each disc
(c) the heat generated per second.

Total braking force per disc

$$F_f = n \times \mu \times p \times A\,[\text{N}]$$

The cross-sectional area of the cylinder, i.e. a circle, is $\pi d^2/4$.

$$F_f = 2 \times 0.32 \times 600 \times 10^3 \left[\frac{\text{N}}{\text{m}^2}\right] \times \frac{\pi \times 0.05^2}{4}\,[\text{m}^2]$$

$$= \mathbf{753.982\ N}$$

$$T = n \times \mu \times p \times A \times r\,[\text{Nm}]$$

$$= 753.982\,[\text{N}] \times 0.15\,[\text{m}]$$

$$= \mathbf{113.097\ Nm}$$

Heat generated per second:

$$\text{power} = T\,[\text{Nm}] \times \omega\,[\text{rad/s}]$$

$$= 113.097\,[\text{Nm}] \times 250 \left[\frac{\text{rev}}{\text{min}}\right] \times \frac{2\pi}{60} \left[\frac{\text{rad/s}}{\text{rev/min}}\right]$$

$$= 2960.872\ \text{Nm/s}$$

$$= \mathbf{2.961\ kJ}\ \text{of heat generated per second}$$

4.5 Centripetal and centrifugal forces

Introduction

Newton's first law of motion states that a body will remain at rest or continue in a straight line with uniform motion unless it is acted upon by an external force. If you were to tie a large nut to a piece of string and spin it around your head, the nut would move in a circular path. The string provides an inward pull to keep the nut moving in the circular path. Let go of the string and the nut will fly off in a straight line. When a body travels in a circular path like this, the outward force that the body applies in attempting to travel in a straight line is called a **centrifugal force.** The inward pull of the string to maintain the circular path is called the **centripetal force**, and is equal and opposite to the centrifugal force (see Newton's third law of motion). Centripetal and centrifugal forces occur whenever a body moves in a circular path. We have established that a body will accelerate if the velocity changes. The velocity will change if either its speed or direction changes. It is the direction that changes when a body moves in a circular path.

Centripetal acceleration

Consider a body moving with uniform speed v in a circular path of radius r. See Figure 4.29.

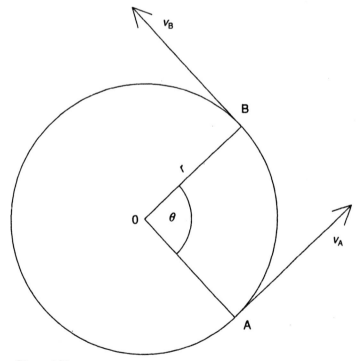

Figure 4.29

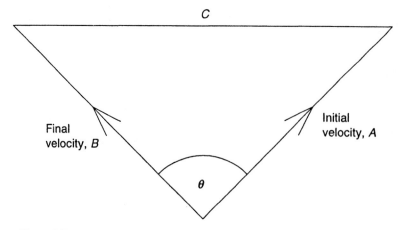

Figure 4.30

The body moves a short distance from A to B in time t. The velocity at A is v and acts tangential to the circle at A. The velocity at B is v and acts tangential to the circle at B. The angular displacement is θ in radians. Although the speed is constant the velocity diagram shows that there is a velocity change, C (see Figure 4.30).

This change is normal to the arc AB. The length of the line C can be estimated by thinking of the triangle ABC as a part of a circle. Consider C to be the arc of a circle. Its length can then be calculated from the angle θ and the radius, v. If the angle θ is small then we can say that:

change in velocity $C = v \times \theta$

(as $S = r\theta$ then $C \approx v\theta$). This is accurate provided that the angle θ is small.

The rate of change of velocity, i.e. acceleration, is equal to:

$$\frac{\text{change in velocity}}{\text{time}}$$

As speed $= \dfrac{\text{distance}}{\text{time}}$, then time $= \dfrac{\text{distance}}{\text{speed}}$

The distance moved through $= r\theta$. The speed is v.

$$\therefore \text{time} = \frac{\text{distance}}{\text{speed}} = \frac{r\theta}{v}$$

$$\text{Acceleration} = \frac{\text{change in velocity}}{\text{time}}$$

$$= v\theta \times \frac{v}{r\theta} = \frac{v^2}{r}$$

As $v = \omega r$ then also acceleration $= \omega^2 r$

This is special type of acceleration. When a body travels at a constant speed in a circular path, the acceleration is due to the constantly changing velocity which is due to the constantly changing direction. The acceleration acts along the radius towards the centre of the circle, O. This is called the **centripetal acceleration.**

Centripetal force

We know from Newton's laws of motion that a body cannot accelerate unless a force is applied to cause that acceleration. For the centripetal acceleration to occur, a **centripetal force** must be applied. This acts in the same direction as the acceleration of course, which is towards the centre of the circle (Figure 4.31).

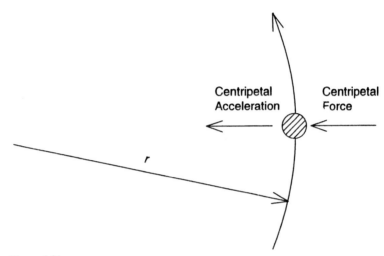

Centripetal
Acceleration

Centripetal
Force

r

Figure 4.31

Centripetal force = mass × centripetal acceleration

$$= m \times \frac{v^2}{r} \quad \text{or} \quad m \times \omega^2 r$$

Units check:

$$[N] = [kg]\left[\frac{m^2}{s^2}\right]\left[\frac{1}{m}\right] = \left[\frac{kg\ m}{s^2}\right]$$

$$\text{or} \quad [N] = [kg]\left[\frac{rad^2}{s^2}\right][m] = \left[\frac{kg\ m}{s^2}\right]$$

(remember that radians do not count when checking units and that 1 N = 1 kg m/s²).

Centrifugal forces

The centripetal force can be applied in various ways. A simple example is someone holding a bucket of water and spinning it around at arm's length. The tension in the person's arm and in the bucket handle provides the centripetal force. The inward pull keeps the bucket on its circular path.

According to Newton's third law there must be an equal force in opposition to the centripetal force. This opposing force is called **centrifugal force** and as it acts radially outwards. This centrifugal force is not applied to the bucket as the centripetal force is; it is applied by the bucket on the handle and the holder's arm. The magnitude of the centrifugal force is equal to the centripetal force, $m\omega^2 r$, but acting in the opposite direction.

Consider a bike turning a corner at speed; centripetal forces are applied to the vehicle from frictional effects between the road and tyres. Centrifugal forces are applied by the bike opposing the centripetal force. If the road and tyre conditions are poor and cannot create enough centripetal force from friction, then there is not enough pull to keep the bike on its circular path and it will skid.

❏ *Example 4.18*

Calculate the centripetal and centrifugal forces that occur when a car takes a corner at 45 km/h. The radius of curvature is 40 m; the mass of the car is 1150 kg; and the driver has a mass of 70 kg. Calculate the centripetal acceleration and the centrifugal force.

$$\text{Centripetal acceleration} = \frac{v^2}{r}$$

$$\text{velocity, } v = 45\left[\frac{km}{h}\right] \times 1000\left[\frac{m}{km}\right] \times \frac{1}{3600}\left[\frac{h}{s}\right] = 12.5\left[\frac{m}{s}\right]$$

$$\text{centripetal acceleration} = \frac{v^2}{r} = \frac{(12.5)^2}{40}\left[\frac{m^2}{s^2}\right]\left[\frac{1}{m}\right] = \textbf{3.906 m/s}^2$$

(acting towards the centre of the arc).

$$\text{Centripetal force} = \text{mass} \times \text{centripetal acceleration}$$

$$= (1150 + 70)\,[kg] \times 3.906\left[\frac{m}{s^2}\right] = \textbf{4765.320 N}$$

As centripetal force magnitude = centrifugal force magnitude then:

centrifugal force = **4765.320 N** acting radially outwards.

Vehicle side skidding and stability

Consider a vehicle negotiating a bend in the road. The vehicle moves in a circular path. The friction forces, within limits, prevent sideways skidding. If the weight of the vehicle mass m is mg and the coefficient of frictional resistance is μ, then the maximum horizontal frictional resistance to sideways skidding is μmg (see page 137). This friction force is the 'pull' to keep the vehicle on its circular path, i.e. the centripetal force. If the force of the vehicle acting on the road, i.e. the centrifugal force mv^2/r, is greater than the centripetal force μmg, then the vehicle will skid. See Figure 4.32.

$$\text{For maximum velocity, } m\frac{v^2}{r} = \mu mg$$

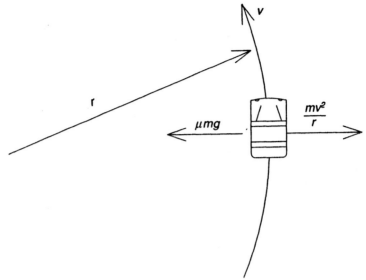

Figure 4.32

This gives $\dfrac{v^2}{r} = \mu g$

Therefore $v^2 = \mu g r$

and $v = \sqrt{\mu g r}$

While the velocity of the vehicle is less than this value, $\sqrt{\mu g r}$, on the particular bend the centrifugal force is less than the maximum possible centripetal force.

If the maximum velocity possible without skidding, $v_{max} = \sqrt{\mu g r}$, then this çan be rearranged to give a minimum safe radius without skidding:

$$\text{minimum safe radius, } r_{min} = \frac{v^2}{\mu g}$$

☐ *Example 4.19*

A car of mass 1320 kg travels round the same level bend every day. The radius of the bend is 32 m. The tyres are new and the coefficient of friction between the tyre and the road is 0.7. Two years later the tyres have worn and the coefficient of friction between the tyre and the road is 0.38. Calculate the maximum velocity the car can take the bend without skidding, when the tyres are new and when they are two years old.

New tyres,

$$v_{max} = \sqrt{\mu g r} = \sqrt{0.7 \times 9.81 \left[\frac{m}{s^2}\right] \times 32\,[m]} = 14.824 \text{ m/s}$$

$$14.824 \left[\frac{m}{s}\right] \times \frac{1}{1000} \left[\frac{km}{m}\right] \times 3600 \left[\frac{s}{h}\right] = 53.366 \text{ km/h}$$

Two years later,

$$v_{\text{max}} = \sqrt{\mu g r} = \sqrt{0.38 \times 9.81 \left[\frac{m}{s^2}\right] \times 32 \, [m]} = 10.922 \text{ m/s}$$

$$10.922 \left[\frac{m}{s}\right] \times \frac{1}{1000} \left[\frac{km}{m}\right] \times 3600 \left[\frac{s}{h}\right] = \textbf{39.319 km/h}$$

When a vehicle is at rest the weight is evenly distributed between the two nearside and two offside wheels. When a vehicle travels round a bend, the weight distribution will change, although the weight of the vehicle will always equal the sum of the reactions of the wheels. If the frictional force preventing sideways skidding is sufficient when a vehicle takes a bend at high speed, then it is possible for the vehicle to overturn. The vehicle tilts about the wheels, B, that are furthest from the centre of the circular path (see Figure 4.33).

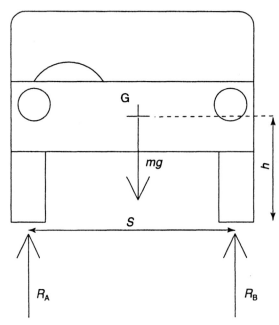

Figure 4.33

    The vehicle is on the point of overturning when the reaction force R_A on the inner wheels is zero and the reaction on the outer wheels R_B is equal to the vehicle weight, *mg*. The centrifugal force of the vehicle acts horizontally through the centre of gravity G. The weight of the vehicle acts vertically through the centre of gravity G. The height of G above the ground is *h*. To consider the stability of a cornering vehicle we can calculate the turning moment about the outer wheels. We assume that the vehicle is symmetrical and that G is positioned at half the distance between the wheels *S*.

The stabilising moment (anti-clockwise)

$$= \text{weight} \times \frac{1}{2} \text{ distance between the wheels}$$

$$= mg \times \frac{1}{2} S$$

The overturning moment (clockwise)

$$= \text{centrifugal force} \times h = m\frac{v^2}{r} \times h$$

At the point of overturning:
overturning moment = stabilising moment

$$m\frac{v^2}{r} \times h = mg \times \frac{1}{2} S$$

$$\frac{v^2}{r} \times h = g \times \frac{1}{2} S$$

The maximum velocity without overturning:

$$(v_{max})^2 = \frac{Sgr}{2h}$$

$$\Rightarrow v_{max} = \sqrt{\frac{Sgr}{2h}}$$

We can rearrange this to give a minimum safe radius without skidding:

$$\frac{v^2 \times h}{r} = g \times \frac{1}{2} S$$

$$\Rightarrow \text{minimum safe radius, } r_{min} = \frac{2v^2 h}{Sg}$$

For this maximum velocity and minimum safe radius for stability, assume that the frictional force between the tyres and the road is sufficient to keep the vehicle on its circular path. If not, then the vehicle will skid before it overturns. The formulae are useful when considering the design of vehicles; the greater the track width S and the smaller the height of the centre of gravity h, then the more stable the vehicle will be on cornering. Notice that racing cars and sports cars tend to be wide and low.

❏ Example 4.20

The track width of a vehicle is 1.62 m. The centre of gravity is 0.8 m above the ground. The coefficient of friction between the tyre and the road is 0.67. The mass of the vehicle is 960 kg. Calculate the maximum speed with which the vehicle can take a bend of radius 27 m without overturning or skidding.

Maximum velocity without skidding:

$$v = \sqrt{\mu g r}$$

$$= \sqrt{0.67 \times 9.81 \left[\frac{m}{s^2}\right] \times 27 \, [m]}$$

$$\doteq 13.322 \ \text{m/s}$$

Maximum velocity without overturning:

$$v = \sqrt{\frac{Sgr}{2h}}$$

$$= \sqrt{\frac{1.62 \, [m] \times 9.81 \, [m/s^2] \times 27 \, [m]}{2 \times 0.8 \, [m]}}$$

$$= 16.376 \ \text{m/s}$$

$$\text{Maximum velocity} = 13.322 \left[\frac{m}{s}\right] \times \frac{1}{1000} \left[\frac{km}{m}\right] \times 3600 \left[\frac{s}{h}\right]$$

$$= 47.959 \ \text{km/h}$$

If this velocity is exceeded then the vehicle will skid.

Bike skidding and stability on cornering

A bike does not have two lines of wheels, only one, making the width of the track S equal to zero. Therefore the formulae for stability previously derived cannot be applied to bikes. There are several forces involved when a bike negotiates a bend. See Figure 4.34.

The centripetal force to 'pull' the bike round the bend is μmg and the reaction of the road on the wheel is R. The weight of the bike acts vertically down through the centre of gravity, G, and the centrifugal force, mv^2/r, acts horizontally outwards through the centre of gravity, G. The height of the centre of gravity above the ground can be varied by the bike leaning over through different angles. The bike pivots about the contact point between the tyre and the road. The distance from the centre of gravity to the pivot point remains constant: l. As the force of gravity acts vertically downwards, then the component of this that causes a clockwise turning moment is $mg \sin \theta$ (see Figure 4.35).

Therefore clockwise turning moment $= mg \sin \theta \times l$. The component of the centrifugal force that causes an anti-clockwise turning moment is $\dfrac{mv^2}{r} \cos \theta$ (see Figure 4.36).

Therefore anticlockwise turning moment $= m \dfrac{v^2}{r} \cos \theta \times l$

For equilibrium,

clockwise turning moment = anticlockwise turning moment

$$\Rightarrow mg \sin \theta \times l = \frac{mv^2}{r} \cos \theta \times l$$

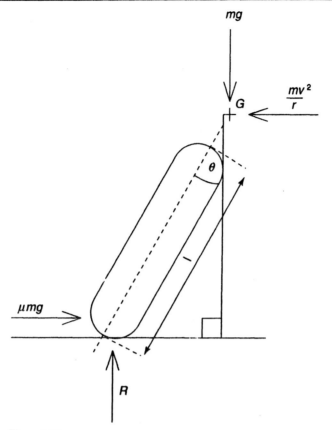

Figure 4.34

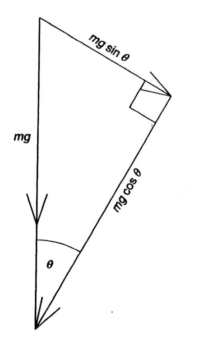

Figure 4.35

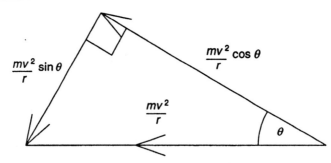

Figure 4.36

$$\therefore v^2 = \frac{rg \sin \theta}{\cos \theta} = rg \tan \theta$$

(as the sin of an angle divided by the cosine of the same angle is equal to the tan of that angle).

$$\therefore v = \sqrt{rg \tan \theta}$$

The formula for the maximum velocity without slipping is the same as for other vehicles, i.e. $v_{\max} = \sqrt{\mu gr}$.

☐ Example 4.21

A bike negotiates a level bend of radius 32 m at a speed of 70 km/h. The coefficient of friction between the tyres and the road is 0.7. What angle must the bike lean over to take that corner? What might prevent the bike from actually taking the corner at this speed?

$$70 \left[\frac{km}{h}\right] \times 1000 \left[\frac{m}{km}\right] \times \frac{1}{36000} \left[\frac{h}{s}\right] = 19.444 \left[\frac{m}{s}\right]$$

Since $v = \sqrt{rg \tan \theta}$

$$\tan \theta = \frac{v^2}{rg}$$

$$\therefore \theta = \tan^{-1} \left(\frac{v^2}{rg}\right)$$

$$= \tan^{-1} \left(\frac{19.444^2 \left[\frac{m^2}{s^2}\right]}{32\,[m] \times 9.81 \left[\frac{m}{s^2}\right]} \right)$$

$$= \tan^{-1}(1.204)$$

$$= \mathbf{50.3°} \quad \text{from the vertical}$$

Obviously there is a limit to how far a bike can lean over before something scrapes the ground, such as the exhaust. Racing bikers can be seen taking bends at very high speeds by leaning the bike over as far as possible and then shifting their body weight further off the seat. This shifts the centre of gravity inwards to the bend centre, increasing the effective angle, θ.

Balancing of machines

If a mass rotates and the centre of gravity is not in the same place as the axis of rotation, then a centrifugal force acts outwards. This happens when a shaft has a mass fastened to it or a wheel is not correctly balanced. This **eccentric load** is very damaging to components and produces excessive vibration and bearing wear. To balance the shaft a balancing mass can be placed diametrically opposite in the plane of rotation. The centrifugal force of the balancing mass must be equal and opposite to the centrifugal force of the eccentric mass. If M_e is the eccentric mass at a radius R from the centre of rotation, then the centrifugal force F_c is:

$$F_c = M_e \times \omega^2 \times R$$

To provide an equal and opposite centrifugal force, a mass; M_b must be placed opposite at a radius, r, as in Figure 4.37.

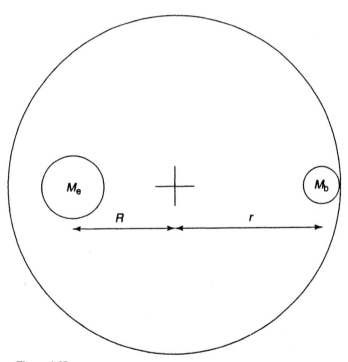

Figure 4.37

To balance $M_e\omega^2 R = M_b\omega^2 r$
$$M_e R = M_b r$$

The product of $M_e R$ or $M_b r$ is the moment of mass sometimes referred to as the **mass moment**. The centre of gravity of the mass must lie in the plane of rotation, or a rocking action will occur. In the chapter on forces we established that, if several co-planar forces are represented by a force polygon, then the polygon must close for equilibrium. If there are several out of balance masses rotating on a shaft at different radii, all in the plane of rotation, they can be balanced in two ways, either by arranging them in such angular positions that their vector diagram closes, or by placing an additional mass so that the vector of its centrifugal force closes the diagram.

□ Example 4.22

Two masses, A and B, are attached to a flywheel. Mass A is 0.5 kg and is placed at a radius of 0.11 m. Mass B is 1.2 kg and is placed at a radius of 0.13 m and at 90° from mass A.

A further mass, C, of 0.75 kg is available to balance the flywheel. Where must it be placed? See Figure 4.38.

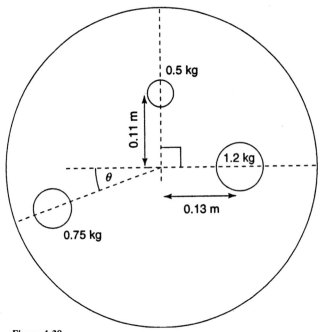

Figure 4.38

Mass moment of A = 0.5 [kg] × 0.11 [m] = 0.055 kg m
Mass moment of B = 1.2 [kg] × 0.13 [m] = 0.156 kg m

Figure 4.39 shows the vector diagram. According to the scale, the mass moment of C is 0.165 kg m.

Mass moment of $C = m_c \times r_c$

$$\therefore r_c = \frac{\text{mass moment of C}}{m_c}$$

$$= \frac{0.165 \,[\text{kg m}]}{0.75 \,[\text{kg}]} = \textbf{0.221 m}$$

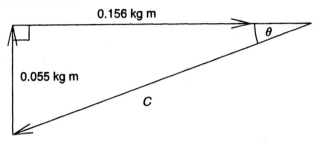

0.156 kg m

0.055 kg m

C

θ

Figure 4.39

$$\tan \theta = \frac{0.055 \, [\text{kg m}]}{0.156 \, [\text{kg m}]}$$

$$\therefore \; \theta = \tan^{-1}\left(\frac{0.055}{0.156}\right) = 19.4°$$

$90° + 19.4° = \mathbf{109.4°}$ anti-clockwise to mass A

A vehicle tyre is manufactured to be balanced but when the complete wheel is assembled, it will often require the addition of a balancing mass. The wheel can be balanced in the plane of rotation, as we have seen, by clipping a single mass on to the rim of the wheel in the correct position. When a wheel requires balancing in this manner it is referred to as a **static unbalance**. If the wheel is spun round with a static unbalance, it will always come to rest in the same position due to the 'out of balance' mass stopping below the axis. If a wheel is used with static unbalance the tyre will wear unevenly.

   Dynamic unbalance occurs when masses creating centrifugal forces do not act in the same plane. This is sometimes referred to as wheel shimmy and causes serious damage. See Figure 4.40.

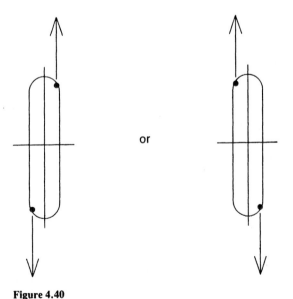

or

Figure 4.40

Static unbalance can be corrected using simple equipment with the wheel stationary and clipping a mass to the wheel rim. More complex equipment is needed to correct dynamic unbalance as the wheel must be rotated to correct the problem.

4.6 Simple harmonic motion

Introduction

Simple harmonic motion is a precise type of periodic motion. Many types of motion in engineering can be considered as being approximately the same as simple harmonic motion which enables calculations to be carried out. For this chapter we will abbreviate simple harmonic motion to SHM.

Defining simple harmonic motion

Think of an engine piston and crank mechanism, as shown in Figure 4.41.

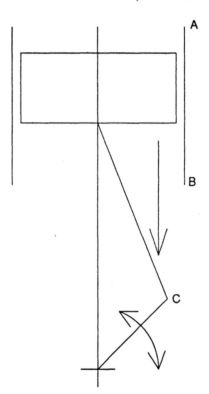

Figure 4.41

The piston moves periodically between A and B. The velocity and acceleration of the piston are not uniform and the formulae we have looked at that deal with motion cannot be used. As the piston moves down and approaches B, its velocity reduces. At B, the piston comes instantaneously to rest before reversing direction. The piston then accelerates from rest and the velocity increases. After the piston has passed the mid-point between A and B, the piston velocity reduces and comes to rest momentarily at A. Then the piston accelerates from rest towards B again and the velocity increases. After the piston has passed the mid-point between A and B the velocity reduces as the piston approaches B again. The process is repeated. As the piston comes to rest momentarily and the direction reverses, the acceleration is at a maximum and the velocity is zero. The velocity increases towards the mid-point and is at a maximum at the mid-point. The acceleration at the mid-point is zero. The motion of a piston can be described approximately by SHM if the crank moves with constant angular velocity. SHM is defined as a periodic motion in which:

1. the acceleration is always directed towards a fixed point in its path
2. the acceleration is proportional to its displacement from a fixed point in its path.

Displacement, velocity and acceleration of simple harmonic motion

In Figure 4.42, the point D moves with SHM along the circle diameter between points A and B. The point C moves with constant angular velocity ω around the circle with radius R and centre O. The motion of C is projected onto the diameter AB.

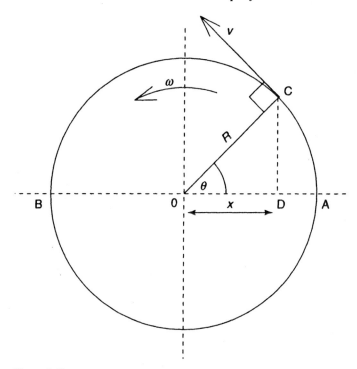

Figure 4.42

Think of an object spinning in a circle in the air in a vertical plane that leaves a shadow on the ground. It is this projected point that is D, moving with SHM. From this we can derive equations of motion for displacement x, velocity v and for acceleration a. The displacement of D from O is x. We can say that:

$$\cos \theta = \frac{x}{R}$$

$$\therefore x = R \cos \theta$$

As $\omega = \theta/t$ then $\theta = \omega t$

then $\therefore x = R \cos \omega t$

The linear velocity of C, $v = \omega R$ around the circular path. A velocity diagram is drawn for the linear velocity of C when the angular displacement is θ as shown in Figure 4.43.

The horizontal component of this diagram represents the velocity of D at that instant. Therefore:

$$\sin \theta = \frac{\text{velocity of D}}{\text{velocity of C}}$$

$\therefore$ velocity of D = velocity of C $\times \sin \theta$

$$= \omega R \sin \theta \text{ m/s}$$

The displacement of D from the mid-point is x. The distance between D and C in triangle OCD can be found from:

$$R^2 = x^2 + DC^2$$

$$\therefore \quad DC = \sqrt{R^2 - x^2}$$

Also $\sin \theta = \dfrac{DC}{R} = \dfrac{\sqrt{R^2 - x^2}}{R}$

Velocity of D $= \omega R \sin \theta$

$$= \omega R \frac{\sqrt{R^2 - x^2}}{R}$$

$$\therefore v = \omega \sqrt{R^2 - x^2}$$

The maximum velocity will occur when the displacement x is zero.

$$v(\text{max}) = \omega R$$

i.e. when $\theta = 90°$ or $270°$.

The centripetal acceleration of a body is equal to $\omega^2 r$ acting towards the centre of rotation. The acceleration of C can be regarded as centripetal acceleration, always acting towards O. Therefore we can say that the acceleration of C $= \omega^2 r$. Another vector diagram may be drawn of the acceleration of C to give:

$$\cos \theta = \frac{\text{acceleration of D}}{\text{acceleration of C}}$$

$\therefore$ Acceleration of D = acceleration of C $\times \cos \theta$

$$= \omega^2 r \cos \theta \text{ m/s}^2$$

Also we can say that $\cos \theta = \dfrac{x}{R}$

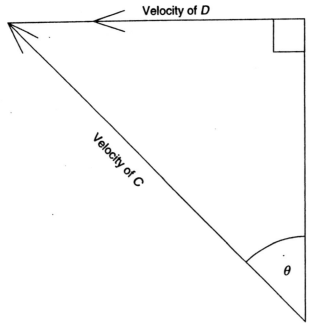

Figure 4.43

$\therefore$ Acceleration of $\mathrm{D} = \omega^2 R \dfrac{x}{R} = \omega^2 x$

$$= \omega^2 \times \text{displacement from mid-point of travel}$$

This agrees with one of the definitions of SHM, that the acceleration is proportional to the displacement x from a fixed point. The maximum displacement x is the radius R and so the maximum acceleration occurs when $x = \pm R$.

$a(\text{max}) = \pm \omega^2 R$

i.e. when $\theta = 0°$ or $180°$.

So to summarise:

$$\text{displacement} = R \cos \omega t$$
$$\text{velocity} = \omega \sqrt{R^2 - x^2}$$
$$\text{maximum velocity} = \omega R$$
$$\text{acceleration} = \omega^2 x$$
$$\text{maximum acceleration} = \omega^2 R$$

Some definitions

Amplitude
The maximum displacement from the centre O is called the amplitude of the motion. If there is no friction, or damping (see later), the amplitude R remains constant.

Periodic time

The periodic time is the time for one complete oscillation or vibration of D, or one complete revolution of C. If the velocity of C is $R\omega$ in metres per second and the displacement for one revolution is $2\pi R$, then the time taken for one revolution is:

$$\text{periodic time} = \frac{\text{displacement}}{\text{velocity}} = \frac{2\pi R}{R\omega}\left[\frac{\text{rad}}{\text{rad/s}}\right]$$

$$= \frac{2\pi}{\omega}\,\text{s}$$

Frequency

The number of complete oscillations, or cycles, occurring in a unit of time is called the frequency f. This is always the reciprocal of periodic time:

$$\text{frequency, } f = \frac{1}{t} = \frac{\omega}{2\pi}$$

The unit of frequency is the hertz [Hz], which is one cycle per second. As acceleration, $a = \omega^2 x$, then $\omega = \sqrt{a/x}$.

$$\text{Periodic time, } t = \frac{2\pi}{\omega} = \frac{2\pi}{\sqrt{a/x}} = 2\pi\sqrt{\frac{x}{a}}\,\text{s}$$

$$\text{Frequency, } f = \frac{1}{t} = \frac{1}{2\pi}\sqrt{\frac{a}{x}}\,\text{Hz}$$

The frequency of an SHM system is also sometimes referred to as the natural phase rate, natural frequency or resonant frequency.

☐ *Example 4.23*

A bike engine revolves with a constant velocity at 1200 rev/min. The speed of a piston crown at a distance of 25 mm from the centre of oscillation is one half of the maximum speed. What is the amplitude and maximum acceleration of the piston crown? Assume that the pistons move with SHM.

The angular velocity of the engine measured in revolutions per second will equal the frequency of the periodic motion of the pistons.

$$1200\left[\frac{\text{rev}}{\text{min}}\right] = 1200\left[\frac{\text{rev}}{\text{min}}\right] \times \frac{1}{60}\left[\frac{\text{min}}{\text{s}}\right] = 20\left[\frac{\text{rev}}{\text{s}}\right] = f$$

As $f = 1/t$ then $t = 1/f$. Periodic time = 1/20 s = 0.05 s.

Velocity, $v = \omega\sqrt{R^2 - x^2}$ and maximum velocity = ωR.

There are a few unknown values at the moment. We know the displacement x when the velocity v is half of the maximum velocity. Write down what you know:

$$v(\text{max})/2 = \omega\sqrt{R^2 - x^2} \tag{1}$$

$$v(\text{max}) = \omega \times R \tag{2}$$

Now substitute equation 2, for $v(\text{max})$ into equation 1:

$$\frac{\omega R}{2} = \omega\sqrt{R^2 - x^2}$$

As the ω value is a multiple on each side, then it can be cancelled from each side. The term on the right has a square root. The formula will be easier to deal with if the left hand term is squared instead:

$$\left(\frac{R}{2}\right)^2 = R^2 - x^2$$

$$R^2 - \frac{R^2}{4} = 0.75R^2 = x^2$$

$$\therefore R = \sqrt{\frac{x^2}{0.75}} = \sqrt{\frac{0.025^2}{0.75}}\,[\text{m}] = 28.867 \times 10^{-3}\,\text{m}$$

The amplitude R will also be the same length as the crank radius causing the reciprocating motion $R = 28.9$ mm.

Maximum acceleration $= \omega^2 R$

$$\left(20\left[\frac{\text{rev}}{\text{s}}\right] \times 2\pi\left[\frac{\text{rad}}{\text{rev}}\right]\right)^2 \times 28.867 \times 10^{-3}\,[\text{m}] = 455.8\,\text{m/s}^2$$

Remember that ω is always measured in radians per second. The units in the above calculation result in $\left[\frac{\text{rad}^2}{\text{s}^2}\right][\text{m}] = \left[\frac{\text{m}}{\text{s}^2}\right]$.

Vibrating mass on a spring

An interesting and useful application of the theory we have derived above is to a mass that vibrates on a helical spring. See Figure 4.44.

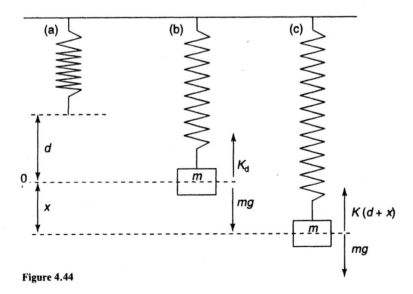

Figure 4.44

A helical spring hangs vertically and carries a mass, m. To simplify things, we will assume that the spring is perfectly elastic and obeys Hooke's law, and also that the mass of the spring is very small compared with the mass m. The force that a spring can exert is calculated using the **spring stiffness**, k, which has units of newtons per metre. The restoring force can then be calculated by multiplying the spring stiffness by the length that the spring is stretched or squashed along its length, known as the deflection d. For example, if a spring has a stiffness k of 200 N/m and when hung vertically supports a mass of 3 kg, then it will deflect as follows:

force in spring = gravitational force on mass

$$k \times d = m \times g$$

$$\therefore d = \frac{mg}{k} = \frac{3 \times 9.81}{200} \, [\text{kg}]\left[\frac{\text{m}}{\text{s}^2}\right]\left[\frac{\text{m}}{\text{N}}\right]$$

$$= 147 \times 10^{-3} \, \text{m} = 147 \, \text{mm}$$

In Figure 4.44, the difference between the spring unstretched and stretched by the mass is d. Imagine that the mass is then pulled down a little further, a distance of x, and let go. The mass would start to oscillate up and down in a way we could describe by SHM. As the mass is released the body moves upwards with an acceleration towards 0. This acceleration can be calculated by adding up the different forces:

total downward force on the spring = mg

total extension of the spring = $(d + x)$

total upward force on the spring = $-k(d + x)$

We put a minus sign in front of this term as x increases downwards and the acceleration increases upwards. It is important to establish a convention and stick to it.

Net restoring force on the spring = $-k(d + x) + mg$

But $kd = mg$ so,

net restoring force becomes $-mg - kx + mg = -kx$

Also, using Newton's second law,

force causing acceleration = mass $\times$ acceleration, ma

So, $ma = -kx$, and $a = \dfrac{-kx}{m}$

This indicates a mass moving with SHM as:

acceleration = constant $\times$ displacement

We can now work out a formula for the frequency of the vibrating mass:

$$\text{frequency}, f = \frac{1}{2\pi}\sqrt{\frac{\text{acceleration}}{\text{displacement}}} = \frac{1}{2\pi}\sqrt{\frac{a}{x}}$$

Substituting for $a = \dfrac{kx}{m}$:

$$f = \frac{1}{2\pi}\sqrt{\frac{kx}{m}\frac{1}{x}} = \frac{1}{2\pi}\sqrt{\frac{k}{m}} \text{ Hz}$$

From the static deflection state we know that $mg = kd$ and so $g/d = k/m$.

Also then $f = \dfrac{1}{2\pi}\sqrt{\dfrac{g}{d}}$ Hz

Notice that the frequency of oscillation depends only on the static deflection, which is itself dependent upon spring stiffness and mass.

❑ *Example 4.24*

For the spring that has a stiffness k of 200 N/m mentioned previously, calculate the frequency of oscillation for the suspended 3 kg mass vibrating and the frequency and periodic time for a 2 kg mass vibrating.

For a 3 kg mass:

$$\text{frequency, } f = \frac{1}{2\pi}\sqrt{\frac{k}{m}} = \frac{1}{2\pi}\sqrt{\frac{200}{3}} \left[\sqrt{\frac{N}{m\ kg}}\right] = \textbf{1.30 Hz}$$

$$\text{Units check: } \left[\sqrt{\frac{N}{m\ kg}}\right] = \left[\sqrt{\frac{kg\ m}{s^2\ m\ kg}}\right] = \left[\sqrt{\frac{1}{s^2}}\right] = [\text{Hz}]$$

For a 2 kg mass:

$$\text{frequency, } f = \frac{1}{2\pi}\sqrt{\frac{k}{m}} = \frac{1}{2\pi}\sqrt{\frac{200}{2}} \left[\sqrt{\frac{N}{m\ kg}}\right] = \textbf{1.59 Hz}$$

$$\text{Periodic time, } \quad t = \frac{1}{f} = \frac{1}{1.59}[\text{Hz}] = \textbf{0.63 s}$$

Pendulums

Another interesting and useful application of SHM theory is to a swinging pendulum. Figure 4.45 shows a mass hanging on the end of a length of string. The other end of the string is attached to something solid. If the mass is moved sideways and released then it will swing and oscillate.

The downward force on the mass is mg. We can split that up into two forces as shown in Figure 4.46 (see page 170).

The force component $mg \sin\theta$ will produce a clockwise torque T about O. For small angles $\sin\theta = \theta^r$, making the force component $mg\ \theta^r$. Now think back to rotational dynamics on page 129. For the mass m the moment of inertia about O, $I_o = ml^2$.

Applying Newton's second law, $\quad T = I_o\alpha$

$$\Rightarrow \quad -mg\ \theta^r\ l = ml^2\alpha$$

Notice the negative sign with the left term. As θ increases away from OX, torque increases towards OX.

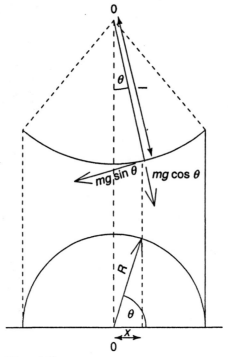

Figure 4.45

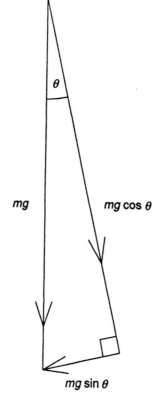

Figure 4.46

$\Rightarrow \quad -g\,\theta^r = l\alpha$

Acceleration, a = angular acceleration × radius

$\therefore a = -g\theta^r$

Provided that the displacement angle θ^r is small then we can say again that: acceleration is proportional to θ^r, as g is constant, and displacement is proportional to θ^r as l is constant. Therefore the acceleration is proportional to displacement and the mass moves with SHM.

We can now work out a formula for the frequency of the swinging pendulum.

$$\text{Frequency, } f = \frac{1}{2\pi}\sqrt{\frac{a}{x}}$$

$$= \frac{1}{2\pi}\sqrt{\frac{g\theta}{l\theta}} = \frac{1}{2\pi}\sqrt{\frac{g}{l}}$$

This shows that the frequency of a pendulum depends upon the length (apart from variation in the value of g) and not on the mass.

☐ Example 4.25

A weight suspended from a length of string forms a simple pendulum. The amplitude, R, is 0.3 m and the periodic time is 4 s. Find the length of the piece of string, and the maximum velocity and maximum acceleration. When the displacement x is 0.1 m, what are the velocity and acceleration?

If frequency, $f = \dfrac{1}{2\pi}\sqrt{\dfrac{g}{l}}$ then rearrange to get:

$$\text{length, } l = \frac{g}{(2\pi f)^2} = \frac{9.81}{(2\pi\frac{1}{4})^2}\left[\frac{m}{s^2}\right][s^2] = 3.976 \text{ m}$$

(remember that $f = 1/t$).

$$\text{Maximum velocity, } V_{max} = \omega R = 2\pi(1/4)0.3\left[\frac{1}{s}\right][m]$$

$$= 0.471 \text{ m/s}$$

$$\text{Maximum acceleration } a_{max} = \omega^2 R = (2\pi(1/4))^2 \times 0.3\left[\frac{1}{s^2}\right][m]$$

$$= 0.741 \text{ [m/s}^2]$$

$$\text{Velocity, } v = \omega\sqrt{R^2 - x^2}$$

$$= 2\pi(1/4)\sqrt{0.3^2 - 0.1^2}\left[\frac{m}{s}\right] = \mathbf{0.444 \text{ m/s}}$$

$$\text{Acceleration, } a = \omega^2 x = (2\pi(1/4))^2\,0.1\left[\frac{1}{s^2}\right][m]$$

$$= \mathbf{0.247 \text{ m/s}^2}$$

□ *Example 4.26*

A pendulum is required to reach the limit of travel each swing every second. Calculate the length of the pendulum.

If it reaches each limit of travel every second then the total periodic time = 2 s.

$$f = \frac{1}{t} = \frac{1}{2\,[\mathrm{s}]} = 0.5 \text{ Hz}$$

$$f = \frac{1}{2\pi}\sqrt{\frac{g}{l}}$$

$$\therefore l = \frac{g}{(2\pi f)^2} = \frac{9.81\left[\frac{\mathrm{m}}{\mathrm{s}^2}\right]}{(2\pi 0.5)^2}$$

$$= 0.994 \text{ m}$$

Damping

So far we have ignored friction with SHM. In practice, friction is always there in some form. A mass vibrating on a spring soon comes to rest if left alone. When energy is drained from an oscillating system, either by friction or some other method, the amplitude of oscillation reduces. This is called **damping**. A system with zero damping would oscillate indefinitely. Extra damping is sometimes caused intentionally in a system. Suspension systems are designed to have damping added to them to limit oscillations of the vehicle. A damper (often called a shock absorber) is fitted to absorb some of the energy stored in the spring and reduce the number of oscillations between a wheel hitting a bump and the spring returning to its rest position.

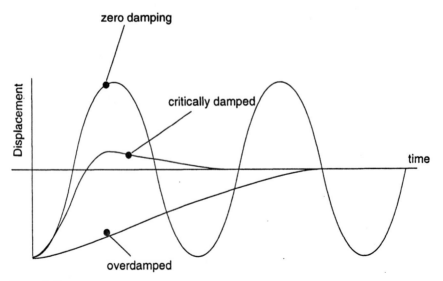

Figure 4.47

Instruments are designed to have damping, so a dial needle does not oscillate but comes to rest in the shortest possible time. When a system just fails to oscillate and comes to rest in the shortest possible time, it is said to be **critically damped**. If it oscillates before coming to rest it is **underdamped**. If it does not oscillate but takes longer to come to rest than if it were critically damped, it is **overdamped**. Figure 4.47 shows a system with a step input applied to it with different amounts of damping applied.

Damping is used to **control** the oscillations of a dynamic system. This is a serious consideration in design.

Problems 4

1. A car moves with a uniform speed of 50 km/h. How long will it take to travel 50 m?
2. A car travels at 50 km/h. It then accelerates at a uniform rate of 2 m/s^2. How long is it before the car reaches a speed of 100 km/h?
3. A car travelling at 60 km/h is involved in an accident and stops in a distance of 20 m. Calculate the rate of retardation and the time taken to stop.
4. A vehicle starts from rest and moves with uniform acceleration for one minute to a speed of 50 km/h. It then continues at this speed for 2 minutes. The vehicle then slows down with uniform retardation to rest over 20 s. Calculate the total distance travelled.
5. The velocity of a lorry is 100 km/h due north. A car travels north-west at 50 km/h. What is the relative velocity of the car to the lorry?
6. A flywheel rotates at a constant angular velocity of 50 rad/s. Calculate the number of revolutions made by a point on the flywheel rim in one minute.
7. The wheels of a vehicle have a diameter of 0.5 m. Calculate the angular velocity of the wheels when the vehicle travels at 100 km/h.
8. A truck of mass 1000 kg travelling at a velocity of 3 m/s due north collides with another truck 1500 kg travelling at a velocity of 1.5 m/s due south. After the impact the trucks remain locked together. Calculate the velocity of the trucks after impact.
9. A vehicle of mass 3000 kg moves at 10 km/h on a level road. It collides with a stationary truck of mass 1500 kg. They move off together after the impact. Calculate the velocity after the impact.
10. A steel component, 0.5 kg, rests on the horizontal surface of a marking table. A force of 0.981 N is required to make the component start to slide over the table surface. The coefficient of kinetic friction between the two surfaces is 0.24. Calculate the coefficient of static friction and the force required to keep the component sliding over the surface at a steady speed.
11. A machine of mass 200 kg is dragged across a horizontal floor. The force required to keep the machine sliding is 510 N. The machine is partially stripped down reducing the mass by 27 kg. What is the new force required to keep the machine moving?
12. A block of mass 20 kg rests on the horizontal ground. The force required to keep the block moving is 47.088 N. What is the angle of friction?
13. The brakes of a car lock all four wheels. The coefficient of friction between the road and the tyres is 0.38. What is the maximum slope that the car can rest on without sliding down the slope?

Hint: look at the section on the inclined plane. In this case though, as the potential motion of the vehicle is down the slope the frictional force acts up the slope opposing the force due to gravity down the slope. The normal reaction R_N is $W \cos \alpha$. The force component acting down the slope due to gravity F is $W \sin \alpha$. As the vehicle just starts to slide the frictional force is equal to this force component acting down the slope. The coefficient of static friction, μ, is therefore equal to $= (F_f/R_N) = (W \sin \alpha)/(W \cos \alpha)$. As the sine of an angle divided by the cosine of that angle is the tangent of that angle, i.e. $(\sin \alpha)/(\cos \alpha) = \tan \alpha$, then the value of $\mu = \tan \alpha$. This is a useful way of determining the coefficient of friction between two surfaces. At the point where a body just about slides down a slope the friction force up the slope equals the gravitational component acting down the slope.

14. Give two examples in a motor vehicle where friction must be kept to a minimum, and two examples where friction is useful.

15. A car has a mass of 1350 kg and travels along a horizontal road. The value of the coefficient of friction μ between the tyres and the road is 0.5. What is the maximum retarding force that can be used without the car skidding?

16. A block with a mass of 20 kg rests on a slope of 10°. The value of the coefficient of friction μ for the two surfaces is 0.25. What force acting parallel to the slope is required to pull the block up the slope?

17. A block with a mass of 15 kg is pulled up a slope of 10°. The value of the coefficient of friction μ is 0.2. The surface of the road changes further up the slope and the value of the coefficient of friction μ changes to 0.26. Calculate the force required acting parallel to the slope before and after the change of road surface.

18. A tool box weighs 18 kg and is dragged along the horizontal ground at a steady speed by a force of 45 N acting at an angle of 10° upwards from the horizontal. What is the coefficient of friction μ?

19. A vice has a spindle with a single start screw thread of mean diameter of 25 mm and pitch 5 mm. It needs to apply a force of 600 N. The coefficient of friction μ between the spindle thread and the housing thread is 0.22. What torque needs to be applied to the spindle?

20. A load of 1000 N is pulled along the ground at a steady speed by a force of 300 N at an angle of 65° to and above the ground. What is the coefficient of friction, μ?

21. A motor vehicle has a single plate clutch with a mean radius of 0.17 m. The friction material has a coefficient of friction of 0.3 and a contact area of $36.5 \times 10^{-3} \, m^2$. The pressure on the friction material is 140 kN/m^2. Calculate the torque transmitted and the power transmitted when the shaft speed is 1500 rev/min.

22. The crank pin of an engine is 0.25 m from the crank shaft centre. The engine rotates uniformly at 2000 rev/min. Calculate the acceleration of the crank pin.

23. A four-wheeled two axle vehicle travels at a uniform speed of 50 km/h around a curve of 100 m radius. The distance between the wheel tracks is 2 m. The centre of gravity of the vehicle is 1.3 m above the ground. The total mass of the vehicle is 3500 kg. Calculate the maximum velocity possible without overturning.

24. A bike negotiates a bend of radius 50 m at a speed of 50 km/h. The coefficient of friction between the tyres and the road surface is 0.6. What angle must the bike

lean over to? What is the maximum speed that the bike can travel around the bend without slipping?

25. Two masses, A and B, are attached to a flywheel. Mass A is 1 kg and attached at a radius of 0.05 m. Mass B is 2 kg and attached at a radius of 0.9 m and at 150° clockwise from 'A'. A further mass of 0.7 kg is available to balance the flywheel. Where must it be placed?

26. Calculate the length of a pendulum that makes four complete oscillations per minute.

27. A piston of mass 0.5 kg moves with SHM at a frequency of 15 Hz. The stroke length is 100 mm. Calculate the acceleration at top dead centre.

5 Work, energy and power

5.1 Work

The definition of work in engineering is rather different from the everyday meaning, and more precise:

> **Work** is said to be done when a force is applied to a body and causes it to move. The quantity of work is measured by multiplying the force by the distance moved in the direction of the force,

> i.e. work = force × distance moved in the direction of force

The units of work are the joule [J]. If you calculate the units you will get [N] × [m] = [Nm]. The definition of the joule is the work done when one newton of force displaces one metre along its line of action. So, 1 Nm = 1 J. If F is the resultant force acting on a body and S is the distance moved in the direction of the force then:

> Work = $F \times S$ [Nm] = FS [J]

❏ *Example 5.1*

Calculate the work done when a car is pushed along a level road in a straight line. The force applied is constant at 350 N and the distance that the car moves is 50 m (Figure 5.1).

$$\text{Work done} = F \times S$$
$$= 350\,[\text{N}] \times 50\,[\text{m}]$$
$$= 17500\,\text{Nm}$$
$$= \mathbf{17.5\,kJ}$$

It is important to remember that only movement in the direction of the force counts towards the work done. This next example will make this clear.

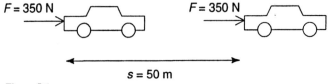

$F = 350\ \text{N}$ $F = 350\ \text{N}$

$s = 50\ \text{m}$

Figure 5.1

□ *Example 5.2*

The car in Example 5.1 is pushed again by a force of 350 N in a straight line along 50 m of level road. This time, however, the force is applied at an angle of 30° to the direction of travel. Calculate the work done (Figure 5.2).

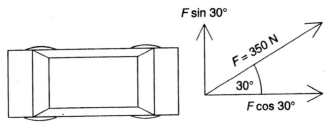

Figure 5.2

The force F can be split into two forces, one $F \sin \theta$ at right angles to the line of travel, the other $F \cos \theta$ acting along the line of travel. The component $F \sin \theta$ does no work at all, as there is no movement along its line of action. The only component in line with the motion of the vehicle is $F \cos \theta$.

Work done = $F \cos \theta \times S = 350 \times \cos 30°$ [N] $\times$ 50 [m]

= 15155.445 Nm = **15.2 kJ**

In this case the force is constant throughout the whole distance. If the force varies uniformly then the average force must be used.

□ *Example 5.3*

A car with mass of 1350 kg is raised 2.5 m on a hydraulic ramp. Calculate the work done.

Work done = $F \times S = m \times g \times S$
= 1350 [kg] $\times$ 9.81 [m/s^2] $\times$ 2.5 [m]
= 33 109 Nm = **33.1 kJ**

If a graph is plotted of force against distance moved, the area under the graph represents the work done.

□ *Example 5.4*

Consider a spring with a stiffness of 10 N/mm. The spring is compressed 10 mm. Calculate the work done.

The first force of 10 N compresses the spring 1 mm, another 10 N (20 N total) compresses the spring another 1 mm (2 mm total) and so on.

$F = kx$ where k is the spring stiffness and x is the distance compressed

$$= 10 \left[\frac{N}{mm}\right] \times 10 \, [mm]$$

$$= 100 \, N$$

$$\text{Average force} = \frac{0 + 100}{2\,[\text{N}]} = 50\,\text{N}$$

Work done = average force × distance
$$= 50\,[\text{N}] \times 10 \times 10^{-3}\,[\text{m}] = \textbf{0.5 J}$$

If a graph is drawn to represent this, it will look like that in Figure 5.3.

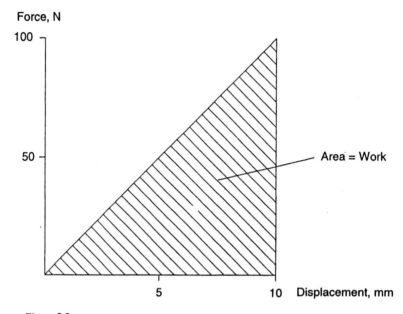

Figure 5.3

Area under graph is a triangle of height 100 N and base 10 mm.

$$\text{Area} = \frac{100\,[\text{N}] \times 10 \times 10^{-3}\,[\text{m}]}{2} = \textbf{0.5 J}$$

A graphical solution can sometimes be much simpler than calculations when the force varies with time. It also provides a clear picture. A work diagram can also be drawn from torque T versus the angle turned through θ for rotating bodies.

5.2 Energy

Energy is defined as the ability to do work. It is the quantity of work stored up in a body. The unit of energy is, therefore, the same as that of work, the joule J. There are many different ways of storing work and so many different forms of energy. The two of interest in mechanics are: **potential energy** and **kinetic energy**.

Potential energy

Potential energy is the energy stored up in a stationary body. This can be due to its position or condition. It is normally abbreviated to PE.

If a body of mass *m* is raised by a vertical height *h* then:

$$\text{work done in raising the mass} = F \times S$$
$$= m \times g \times h$$
$$\text{Potential energy} = mgh$$

Work is done on the mass to change its position. At the height *h* the body has stored up in it *mgh* joules of potential energy. If an engine of weight 1500 N is raised 2 m with a crane, the work done in this lift is 1500 [N] × 2 [m] = 3000 [Nm]. At this height the engine has stored up in it 3 kJ of potential energy, because of its position. If the engine is allowed to fall back to the ground, then it would give up this amount of work.

Potential energy can be stored up in a body because of other conditions. A loaded suspension spring possesses potential energy. Pressurised hydraulic oil in a braking system is another example of potential energy.

❑ Example 5.5

A mass of 1 kg falls from a height of 10 m onto a vertical pile that is to be driven into the ground. The resisting force of the ground on the pile is 7 kN. Ten per cent of the available energy is dissipated on impact as sound, heat, etc. How far is the pile driven into the ground?

$$\text{Potential energy} = mgh$$
$$= 1 \, [\text{kg}] \times 9.81 \left[\frac{\text{m}}{\text{s}^2}\right] \times 10 \, [\text{m}]$$
$$= 98.1 \, \text{Nm}$$

As 10% of the energy is dissipated then,

$$\text{energy available for driving} = 0.9 \times 98.1 \, [\text{Nm}]$$
$$= 88.29 \, \text{Nm}$$

88.29 J of energy are available to do work.

$$\text{Work on pile} = \text{force (resisting)} \times \text{depth (into ground)}$$

$$\text{Therefore depth} = \frac{\text{work}}{\text{force}}$$

$$= \frac{88.29 \, [\text{Nm}]}{7 \times 10^3 \, [\text{N}]}$$

$$= 12.613 \times 10^{-3} \, \text{m} = \textbf{12.6 mm}$$

Kinetic energy

The kinetic energy of a body is the stored energy it possesses due to its motion. This is usually abbreviated to KE. Think of a body mass *m* being pushed by a force that is greater than the resistance forces such as friction. The force F will cause an acceleration in the body. The work done in creating the acceleration will be stored in the body in the form of kinetic energy:

work done = force × distance

force F = mass × acceleration

$$\text{acceleration} = \frac{\text{change in velocity}}{\text{time}} = \frac{v - 0}{t} = \frac{v}{t}$$

$$\therefore F = \frac{m \times v}{t}$$

distance = average velocity × time

$$= \frac{0 + v}{2} \times t = \frac{vt}{2}$$

work done = force × distance

$$= \frac{mv}{t} \times \frac{vt}{2} = \frac{1}{2}mv^2$$

$$\text{kinetic energy} = \frac{1}{2}mv^2$$

The same derivation can be carried out for a force bringing a moving body to rest. In this case, as the change in velocity is negative, the kinetic energy is found to be negative. The kinetic energy of a body with only linear motion is referred to as **kinetic energy of translation**.

If a moving body of mass m has its velocity changed from v_1 to v_2 due to a resultant force F acting over a distance S then:

change in kinetic energy = final KE − initial KE

$$= \frac{1}{2}mv_2^2 - \frac{1}{2}mv_1^2$$

$$= \frac{1}{2}m(v_2^2 - v_1^2)$$

If the 1500 N engine raised to a height of 2 m is allowed to fall it gains kinetic energy as it increases in velocity. The gain in kinetic energy is equal to the loss of potential energy. As the engine reaches the ground it has fallen 2 m, lost all its potential energy and gained an equal amount of kinetic energy.

☐ *Example 5.6*

A car with a mass of 1000 kg accelerates uniformly from rest to a speed of 15 m/s due to a resultant force of 5 kN. Calculate the work done by the accelerating force and the increase in kinetic energy.

Work done = force × distance

We know the force, we need the distance. We could use one of the equations of motion, for instance $v^2 = u^2 + 2aS$, if we knew the acceleration a.

From force = mass × acceleration

$$\text{acceleration} = \frac{\text{force}}{\text{mass}} = \frac{5000\,[\text{N}]}{1000\,[\text{kg}]} = 5 \text{ m/s}^2$$

If $v^2 = u^2 + 2aS$,

$$\text{then } S = \frac{v^2 - u^2}{2a} = \frac{15^2 - 0}{2 \times 5}\left[\frac{\text{m}^2}{\text{s}^2}\right]\left[\frac{\text{s}^2}{\text{m}}\right]$$

$$= 22.5 \text{ m}$$

$$\text{Work done} = F \times S = 5000\,[\text{N}] \times 22.5\,[\text{m}]$$

$$= 112.5 \times 10^3 \text{ Nm} = \mathbf{112.5\,kJ}$$

As the kinetic energy at the start is zero then:

$$\text{increase in KE} = \frac{1}{2} \times 1000\,[\text{kg}] \times 15^2\,[\text{m}^2/\text{s}^2]$$

$$= 112.5 \times 10^3 \text{ Nm} = \mathbf{112.5\,J}$$

When calculating the units, remember that $1 \text{ N} = 1 \text{ kg m/s}^2$ and so $1 \text{ kg} = 1 \text{ N s}^2/\text{m}$. Hence the work done is equal to the increase in kinetic energy. When the engine above is allowed to fall the potential energy is converted into kinetic energy; this brings us on to the next point.

Energy can neither be created or destroyed.

This is the **principle of conservation of energy**. All forms of energy are transferable. The work done in accelerating a body reappears as kinetic energy. The potential energy 'lost' by a falling body reappears as kinetic energy as its velocity increases. All energy can be converted. It is not always converted into a form that is useful. As friction is a resistance force, then work is done in overcoming friction, and energy is converted into heat. Energy is sometimes considered 'lost' when it cannot be converted into a form to do useful work.

Electricity is a form of energy. An electric fire is designed to convert all of its electrical energy into heat. When an electrical appliance does not work properly it can overheat. This is because if the appliance does not work correctly and perform the energy conversions it is designed to do, the excess electrical energy is converted into heat.

Work and energy of a rotating body

Equations on linear motion can be converted into a form for rotational dynamics. The same is true for the kinetic energy formula. The kinetic energy of a rotating body is referred to as the **kinetic energy of rotation**. We know that $\text{KE} = \frac{1}{2}mv^2$ where v is the linear velocity in metres per second. For a rotating body, the whole mass can be considered as acting at a radius called the radius of gyration k. Check back to the chapter on motion if you are not too sure of this bit.

Linear velocity = angular velocity × radius

Remember that the angular velocity is measured in radians per second.

$$v = \omega \times k$$
$$v^2 = \omega^2 \times k^2$$

We can now substitute this back into the formula for KE.

$$\text{Kinetic energy} = \frac{1}{2}mv^2$$

$$\text{kinetic energy} = \frac{1}{2}m(\omega^2 k^2)$$

The moment of inertia for a rotating body I is equal to the mass multiplied by the square of the radius of gyration:

$$I = mk^2$$

This can now be substituted back into the equation:

$$\text{kinetic energy} = \frac{1}{2}m(\omega^2 k^2) = \frac{1}{2}I\omega^2$$

You can check the units:

$$\text{as } 1\,[\text{N}] = 1\left[\frac{\text{kg m}}{\text{s}^2}\right]$$

$$\text{then } [\text{kg m}^2]\left[\frac{\text{rad}}{\text{s}}\right]^2 = [\text{Nm}]$$

The angular equivalent of Newton's second law of motion, $F = ma$, is $T = I\alpha$ (see the section on motion). In a similar way, the work done by an angular system is calculated as follows:

$$\text{Work done} = F\,[\text{N}] \times S\,[\text{m}] \qquad \text{(linear motion)}$$

As $S = r\theta^r$ then

$$\text{Work done} = F\,[\text{N}] \times r\theta\,[\text{m}]$$
$$= T\,[\text{Nm}] \times \theta^{[r]} \qquad \text{(angular motion)}$$

❑ Example 5.7

The bearing surface of a shaft is to be reground. The tool is applied radially and the cutting force remains constant throughout the operation at 1 kN. The diameter of the shaft is 56 mm. The operation is completed in 500 revolutions of the shaft. Calculate the work done.

$$\text{Torque} = \text{cutting force} \times \text{radius}$$
$$T = F \times r$$
$$= 1 \times 10^3\,[\text{N}] \times \frac{56 \times 10^{-3}}{2}\,[\text{m}]$$
$$= 28\,\text{Nm}$$

Rotational work done = torque × angle turned through

$$= T\theta$$

$$= 28\,[\text{Nm}] \times 500\,[\text{rev}] \times 2\pi \left[\frac{\text{rad}}{\text{rev}}\right]$$

$$= 87.965 \times 10^3\,\text{Nm} = \mathbf{87.965\,kJ}$$

☐ *Example 5.8*

A rotating shaft and flywheel have a total moment of inertia of 5 kg m^2 and rotate at 3000 rev/min. If the frictional resistance of the engine is 2 Nm and that value is constant over all speeds, find the number of revolutions the engine will turn before it comes to rest when the fuel is cut.

KE stored in flywheel and shaft $= \tfrac{1}{2}I\omega^2$

$$= \frac{1}{2} \times 5\,[\text{kg m}^2] \times \left(3000\left[\frac{\text{rev}}{\text{min}}\right] \times 2\pi\left[\frac{\text{rad}}{\text{rev}}\right] \times \frac{1}{60}\left[\frac{\text{min}}{\text{s}}\right]\right)^2$$

$$= 246.74 \times 10^3\,\text{Nm}$$

Rotational work done = torque × angle turned through

$$= T\theta$$

Work for one revolution $= 2\,[\text{Nm}] \times 2\pi$

$$= 12.566\,\text{Nm}$$

$$\text{Number of revolutions} = \frac{\text{total loss of KE}}{\text{KE lost per revolution}}$$

$$= \frac{246.74 \times 10^3\,[\text{Nm}]}{12.566\,[\text{Nm/rev}]}$$

$$= \mathbf{19635.5\,rev}$$

5.3 Power

In engineering, the rate at which work can be done is called the **power**. To calculate the power of a machine, divide the total work it does by the time it takes:

$$\text{power} = \frac{\text{work done}}{\text{time taken}}$$

The unit of power is therefore the newton-metre per second (N m/s) which is equal to the joule per second (J/s). This unit is given the title of the **watt** and the symbol W. One watt is equal to one joule per second. The quantity symbol of power is P. A more powerful machine can do the same amount of work as a less powerful machine in less time. Compare two trucks that have to carry the same weight load up a hill. The truck with the most powerful engine will reach the top first. If the more powerful truck reaches the top in half the time then it must be twice as powerful. The old imperial unit of power that you may come across is the horsepower.

1 horsepower = 745.7 W

As an approximate 'rule of thumb' calculation one horsepower equals 0.75 kW.

❏ *Example 5.9*

The mean torque developed by an engine shaft is 407 Nm when running at 3000 rev/min. Calculate the power developed.

$$\text{Rotational speed} = 3000 \left[\frac{\text{rev}}{\text{min}}\right] \times \frac{1}{60}\left[\frac{\text{min}}{\text{s}}\right] \times 2\pi \left[\frac{\text{rad}}{\text{rev}}\right]$$

$$= 314.159 \left[\frac{\text{rad}}{\text{s}}\right]$$

$$\text{power developed} = 314.159 \left[\frac{\text{rad}}{\text{s}}\right] \times 407 \, [\text{Nm}]$$

$$= 127.863 \times 10^3 \, \text{Nm/s} = \mathbf{127.9 \, kW}$$

❏ *Example 5.10*

A car has a total resistance of 1 kN when driven 100 m in 20 s. What is the power developed by the engine?

Work = force × distance moved in the direction of force

$$\text{power} = \frac{\text{force} \times \text{distance moved in the direction of force}}{\text{time taken}}$$

$$= \frac{1 \times 10^3 \, [\text{N}] \times 100 \, [\text{m}]}{20 \, [\text{s}]}$$

$$= \mathbf{5 \, kW}$$

In a similar manner, angular power can be calculated. If angular work is equal to torque multiplied by angular distance:

$$\text{work done} = T \, [\text{Nm}] \times \theta^{[\text{r}]}$$

then angular power is equal to torque multiplied by angular velocity:

$$\text{power} = T \, [\text{Nm}] \times \omega \, [\text{rad/s}]$$

Calculation of engine power

The force F acting on a piston can be calculated from the pressure p on the piston multiplied by the area A:

$$p = \frac{F}{A} \quad \therefore F = pA$$

The pressure varies on the piston throughout the power stroke. In order to calculate the work done on the piston, the average effective pressure is used. This is called the

mean effective pressure. If p_m is the mean effective pressure and A is the piston area then:

average force on the piston, $F = p_m \times A$

If the length of stroke is l then:

work done $\qquad\qquad$ = force × distance
$\qquad\qquad\qquad\qquad\qquad$ = $p_m \times A \times l$

The power developed is the work done per second. If n is the number of power strokes per second, then:

work done per second = $p_m \times A \times l \times n$

$$\text{For a 4-stroke engine, } n = \frac{\text{engine speed} \left[\dfrac{\text{rev}}{\text{s}}\right]}{2}$$

$$\text{For a 2-stroke engine, } n = \text{engine speed} \left[\frac{\text{rev}}{\text{s}}\right]$$

☐ *Example 5.11*

A four stroke engine has a bore of 75 mm and a stroke of 87 mm. The mean effective pressure at a speed of 3000 rev/min is 4.7 MN/m². Calculate the power developed at this speed.

Power = $p_m \times A \times l \times n$

$$= 4.7 \times 10^6 \left[\frac{N}{m^2}\right] \times \frac{\pi(75 \times 10^{-3})^2}{4} \, [m^2] \times 87 \times 10^{-3} \, [m] \times \frac{3000}{60 \times 2} \left[\frac{1}{s}\right]$$

$$= 4.7 \times \pi \times \frac{75^2}{4} \times 87 \times 25 \times 10^{-3} \left[\frac{Nm}{s}\right]$$

$$= 45.162 \times 10^3 \, W = \mathbf{45.1 \, kW}$$

☐ *Example 5.12*

A solid wheel of diameter 0.4 m and mass 4.5 kg rolls along the ground at a velocity of 3 m/s. Find the total kinetic energy. Assume that the radius of gyration, $k = r/\sqrt{2}$.

$$k = \frac{r}{\sqrt{2}} = \frac{0.4 \, [m]}{2 \times \sqrt{2}} = 0.141 \, m$$

Provided that the units of the kinetic energy of rotation and the kinetic energy of translation are the same, then they can be added together.

If $v = \omega r$ then $\omega = \dfrac{v}{r}$

$$\therefore \omega = \frac{3 \, [m/s]}{0.2 \, [m]} = 15 \, rad/s$$

Total KE = KE of translation + KE of rotation

$$= \frac{1}{2}mv^2 + \frac{1}{2}I\omega^2$$

$$= \frac{1}{2}mv^2 + \frac{1}{2}mk^2\omega^2$$

$$= \frac{1}{2} \times 4.5 \times 3^2 + \frac{1}{2} \times 4.5 \times 0.141^2 \times 15^2 \left[\frac{\text{kg m}^2}{\text{s}^2}\right]$$

$$= 20.250\,[\text{Nm}] + 10.065\,[\text{Nm}]$$

$$= \mathbf{30.315\,Nm}$$

A work diagram for an engine of torque versus angle turned can be drawn. The energy produced in one revolution is the work done in 2π rad. The torque of an internal combustion engine varies considerably through a cycle of operation, depending upon the piston force and the crank leverage. The torque reaches a peak during the power stroke. During the exhaust, induction and compression strokes (assuming the engine is a four-stroke one) energy must be supplied to keep the engine running.

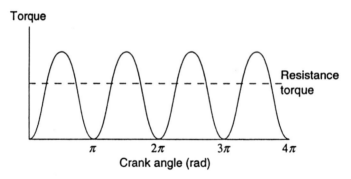

Figure 5.4

Look at Figure 5.4. Each cylinder of a four-stroke engine fires once every two revolutions of the engine. The resistance torque due to loading and friction remains fairly constant. At some parts of the cycle the engine produces more energy than is required (where the line is above the resistance torque line). The excess energy produced is absorbed by the flywheel and stored as kinetic energy. When this happens the engine increases in speed. The flywheel acts as a reservoir of kinetic energy. When more torque is required than is produced at that time by the engine, the engine speed drops briefly and the flywheel gives up some of its energy to the shaft (where the line is below the resistance torque line). The cyclic variation of the engine is greatly reduced by the flywheel and it can keep the engine speed between two limits of operation. The greater the inertia of the flywheel, the less variation in speed occurs. Over one cycle, the energy supplied by the engine must equal the energy required. The transmission system must be designed to transmit the maximum torque, although the average torque is much lower.

5.4 Lifting Machines

The power of a machine, such as a motor vehicle, is used to overcome:

1. the inertia of the load, i.e. accelerating the load
2. effect of gravitation
3. external resistance forces or loads
4. friction of the machine.

When a motor vehicle runs there are many different energy changes taking place at once. The energy source is the fuel flowing from the tank to the engine and this is converted to other forms of energy by the vehicle: the vehicle needs to overcome the resistance to movement; there are changes in the potential energy of the vehicle as it travels up and down hills; heat is transferred to the atmosphere, mainly from the radiator; the energy of the exhaust gases passes to the atmosphere. Also, consider the brakes. When the speed of a vehicle is reduced by applying the brakes, the surface of the brake pads is forced onto the rotating surface of the brake disc to create a frictional force that opposes the relative motion between the two surfaces. The harder the brakes are applied the greater the perpendicular force between the two surfaces and the more frictional resistance is caused. The kinetic energy of the vehicle is therefore converted into heat energy by the braking system and this is transferred to the atmosphere.

The **efficiency** of a machine is the ratio of the output to the input:

$$\text{efficiency} = \frac{\text{energy out}}{\text{energy in}}$$

The energy out is the performance of a machine. The energy in is what is required for the machine to operate, the part we usually have to pay for, e.g. fuel. The higher the efficiency of a machine, the less energy input is required for the same energy output. The difference between the two represents energy 'losses' such as overcoming friction and other resistance forces, and should be kept as low as possible. Internal friction of a machine causes an opposing force and so work is done in overcoming this. The efficiency then is the energy out of a system measured as a fraction of the energy in. Usually this is expressed as a percentage:

$$\text{efficiency} = \frac{\text{energy out}}{\text{energy in}} \times 100\%$$

This can also be calculated from the work done by a machine and the work supplied to the machine.

$$\text{efficiency} = \frac{\text{useful work done}}{\text{work supplied}} \times 100\%$$

Usually though efficiency is calculated in terms of power, i.e.

$$\text{efficiency} = \frac{\text{power in}}{\text{power out}} \times 100\%$$

See Figure 5.5.

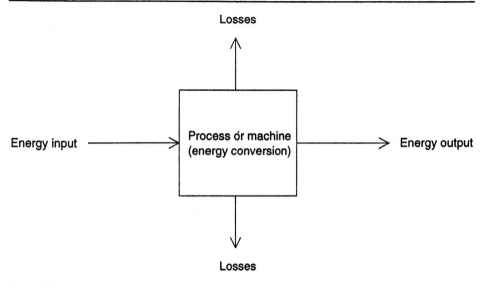

Figure 5.5

A machine is a device that converts energy from one form to another, to make work easier. This usually involves overcoming a resistance force or load by applying another force that is more convenient.

The **mechanical advantage** or **force ratio** of a machine is the ratio of the load to the effort. This applies to all lifting machines.

$$\text{Mechanical advantage} = \frac{\text{load}}{\text{effort}}$$

The units of the load and the effort are the same and so mechanical advantage has no units. It is abbreviated to MA.

However, the distances moved by the load and the effort are not the same. If the load is heavier than the effort, then the effort must be moved through a greater distance than the load will be lifted. The ratio of the distance moved by the effort to the distance moved by the load is called the **velocity ratio**.

$$\text{Velocity ratio} = \frac{\text{distance moved by effort}}{\text{distance moved by load}} = \frac{a}{b}$$

As with mechanical advantage, the units of the distance of the effort and the distance of the load are the same and so velocity ratio has no units. Velocity ratio is abbreviated to VR. Here the mechanical advantage has the same value as the velocity ratio but this assumes that there are no losses due to friction. In practice, a machine always has losses and the velocity ratio is not the same as the mechanical advantage. In fact the efficiency of a machine can be calculated from the mechanical advantage and the velocity ratio. The efficiency of any machine or system can be calculated from the ratio of the useful work done to the work supplied:

$$\text{efficiency} = \frac{\text{useful work done}}{\text{work supplied}} \times 100\%$$

$$= \frac{\text{load} \times \text{distance moved by the load}}{\text{effort} \times \text{distance moved by the effort}} \times 100\%$$

Now, $\dfrac{\text{load}}{\text{effort}} = \text{MA}$ and $\dfrac{\text{distance moved by effort}}{\text{distance moved by load}} = \text{VR}$

so efficiency $= \text{MA} \times \dfrac{1}{\text{VR}} \times 100\% = \dfrac{\text{MA}}{\text{VR}} \times 100\%$

Pulley systems

A pulley is a simple machine. A rope pulley consists of two pulley blocks, one at the top and one at the bottom. Each pulley block has several pulley wheels that can turn freely. There will be either the same number of pulley wheels in each block or there will be one more in one than the other. A rope is threaded over each pulley in turn. One end of the rope is fastened to the block opposite the last pulley. The other end is used to apply the effort. The effort is directed downwards and the load moves up (see Figure 5.6).

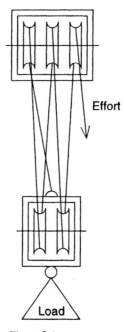

Figure 5.6

When a pulley system is connected like this there are five ropes between the load block. If we want to lift the load by say one metre, then all of the five lengths of the rope must be shortened by one metre. To do this, then the effort must pull down by five metres.

$$\text{Velocity ratio} = \frac{\text{distance moved by effort}}{\text{distance moved by load}} = \frac{5}{1} = 5$$

The same calculation can be applied to any number of pulley wheels:

i.e. VR of rope pulley blocks = number of ropes lifting the load block

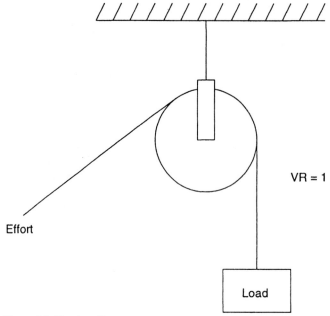

VR = 1

Effort

Load

Figure 5.7 Simple pulley

Also, the number of ropes supporting the load is equal to the number of pulleys in the system. Hence, the number of pulleys is equal to the VR (see Figure 5.7). For example, a system with two pulleys at the top and one at the bottom has a VR of three; a system with three pulleys at the top end and three at the bottom has a VR of six.

This is only true when the direction of the effort opposes the direction of the movement of the load, which is the usual case. If the rope is wound in some way so that the direction of the effort is the same as that of the load then the velocity ratio equals the total number of pulleys plus one.

☐ *Example 5.13*

A rope pulley system has two pulleys in each block. An effort of 116 N is required to lift a load of 390 N. Calculate the efficiency of the machine.

There are four pulleys in total and therefore the VR = 4.

$$\text{Mechanical advantage} = \frac{\text{load}}{\text{effort}}$$

$$= \frac{390\,[\text{N}]}{116\,[\text{N}]} = 3.362$$

$$\text{Efficiency} = \frac{\text{MA}}{\text{VR}} \times 100\%$$

$$= \frac{3.362}{4} \times 100\% = \mathbf{84.05\%}$$

Weston differential pulley block

This type of machine uses three pulleys. The velocity ratio depends upon the diameters of the larger and smaller pulleys. See Figure 5.8.

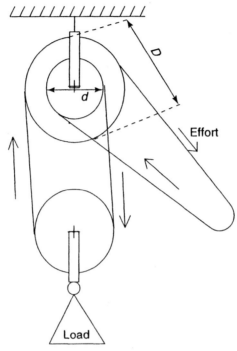

Figure 5.8 Weston differential pulley block

Chains are usually used instead of rope so the pulleys have teeth to take the chain links. When the effort is applied the compound pulley turns. The chain is pulled off the larger pulley by the effort and onto the smaller one. The load chain is pulled off the smaller pulley and onto the larger one. The load moves up due to the different diameters of the compound pulley wheels.

D is the diameter of the larger pulley and if d is the diameter of the smaller pulley, then for one revolution of the compound pulley:

distance moved by effort $= \pi D$ (length of circumference)

distance moved by load $= \frac{1}{2}(\pi D - \pi d)$

$$VR = \frac{\text{distance moved by effort}}{\text{distance moved by load}}$$

$$= \frac{\text{circumference of the big pulley}}{\text{half the difference of the two pulley wheels}}$$

$$= \frac{\pi D}{\frac{1}{2}(\pi D - \pi d)} = \frac{2\pi D}{\pi(D - d)}$$

$$= \frac{2D}{D - d}$$

If the number of teeth is constant then this can be written as:

$$VR = \frac{2 \times \text{number of teeth in larger pulley}}{\text{difference in the number of teeth between pulleys}}$$

☐ Example 5.14

The diameters of the larger and smaller pulleys of a Weston differential pulley block are 105 mm and 95 mm, respectively. A load of 3 kN is lifted with an effort of 238 N. Calculate the efficiency.

$$VR = \frac{2D}{D-d} = \frac{2 \times 105}{105 - 95} = 21$$

We can calculate the VR from [mm] lengths as the units cancel:

$$\text{mechanical advantage} = \frac{\text{load}}{\text{effort}}$$

$$= \frac{3 \times 10^3 \, [\text{N}]}{238 \, [\text{N}]} = 12.6$$

$$\text{efficiency} = \frac{\text{MA}}{\text{VR}} \times 100\%$$

$$= \frac{12.6}{21} \times 100\% = \mathbf{60\%}$$

5.5 Power transmission

The output power of an engine shaft is referred to as the **brake power** or **shaft power**. This is because it is measured by applying a brake to the shaft. The braking force can be measured and from this the torque calculated. By multiplying the speed of the engine by the torque applied, the power output P can be calculated from $P = T\omega$. When power is transmitted by belt of chain from one pulley to another, provided that there is no slipping, the linear velocity of the rim of each pulley must be the same, since they are both driven by the same chain. If the radius of the follower pulley is r_f and the radius of the driver pulley is r_d then using $v = r\omega$:

linear speed of rim of driver = linear speed of rim of follower
$$r_d\omega_d = r_f\omega_f$$

For a chain drive, the number of teeth of the wheel circumference can be used. If n_d is the number of teeth of the driver (sprocket) circumference and n_f is the number of teeth on the follower (chain wheel) circumference:

$$n_d\omega_d = n_f\omega_f$$

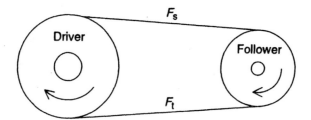

Figure 5.9

Consider the chain drive in Figure 5.9. The chain tension is different on each side of the drive. If F_t is the tension or force on the tight side and F_s is the tension or force on the slack side, then the effective driving force available on the follower is $F_t - F_s$. Therefore the power transmitted,

$$P = (F_t - F_s) \, [N] \times \text{linear speed of the chain} \left[\frac{m}{s}\right]$$

$$= (F_t - F_s) \, [N] \times \omega r \left[\frac{m}{s}\right]$$

$$= (F_t - F_s) \, [N] \times n \left[\frac{rev}{s}\right] \times 2\pi \left[\frac{rad}{rev}\right] \times r \, [m]$$

$$= (F_t - F_s) \times 2\pi n r \text{ Nm/s or W}$$

$$VR = \frac{\text{distance moved by effort}}{\text{distance moved by load}}$$

If we are considering the distance moved in one second,

$$VR = \frac{\text{revolutions per second of driver}}{\text{revolutions per second of follower}}$$

$$= \frac{\omega_d}{\omega_f}$$

Using $r_d \omega_d = r_f \omega_f$

$$\frac{\omega_d}{\omega_f} = \frac{r_f}{r_d}$$

$$\therefore \; VR = \frac{r_f}{r_d}$$

$$\text{or } VR = \frac{D_f}{D_d}$$

$$= \frac{\text{number of teeth of chain wheel}}{\text{number of teeth of sprocket}}$$

Gear wheels

Motor vehicles usually use gear wheels for power transmission. Power is transmitted from one shaft to another, but the teeth are designed to mesh directly. Let T_d and T_f be the number of teeth of the driver and follower respectively. Then,

$$VR = \frac{\omega_d}{\omega_f} = \frac{D_f}{D_d} = \frac{T_f}{T_d}$$

When two gear wheels are used, a driver and a follower, they rotate in opposite directions as in Figure 5.10.

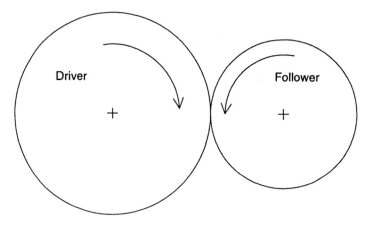

Figure 5.10

If an idler wheel is used between the two wheels, the driver and the driven wheels rotate in the same direction as in Figure 5.11.

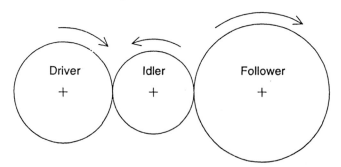

Figure 5.11

The number of teeth on the idler wheel does not affect the VR of the driver and the follower wheel. If ω_d, ω_i and ω_f are the rotational speeds of the driver, idler and follower respectively, and T_d, T_i and T_f are the number of the teeth of the driver, idler and follower respectively, then, consider the driver and the idler:

$$\frac{\omega_d}{\omega_i} = \frac{T_i}{T_d}$$

$$\therefore \omega_i = \frac{T_d \times \omega_d}{T_i} \tag{1}$$

Consider the idler and the follower:

$$\frac{\omega_i}{\omega_f} = \frac{T_f}{T_i}$$

$$\therefore \omega_f = \frac{T_i \times \omega_i}{T_f} \tag{2}$$

Substitute equation for ω_i into equation 2:

$$\omega_i = \frac{T_d \times \omega_d}{T_i}$$

$$\omega_f = \frac{T_d \times \omega_d}{T_i} \times \frac{T_i}{T_f}$$

$$= \frac{T_d \times \omega_d}{T_f}$$

$$\therefore \frac{\omega_d}{\omega_f} = \frac{T_f}{T_d} = \text{VR}$$

This is the same as the previous derivation without the idler wheel. Therefore the velocity ratio between the follower and the driver is independent of the number of teeth on the idler wheel. The only purpose of the idler is to change the direction of rotation of the follower.

The ratios used in gear boxes are achieved by **compound gear trains**. Several trains are used in series, with the driven gear wheel of one train fixed onto the same shaft as the driver of the next. Any idler wheels in the system have no effect on the velocity ratio. All of the gear wheels apart from the idlers can be classed as either drivers or followers. See Figure 5.12.

Consider driver 1 and follower 1:

$$\frac{\omega_{d1}}{\omega_{f1}} = \frac{T_{f1}}{T_{d1}}$$

$$\therefore \omega_{f1} = \frac{T_{d1} \times \omega_{d1}}{T_{f1}}$$

As d_2 is fixed to the same shaft as f1 they must rotate at the same speed.

i.e. $\omega_{f1} = \omega_{d2}$

Consider driver 2 and follower 2:

$$\frac{\omega_{d2}}{\omega_{f2}} = \frac{T_{f2}}{T_{d2}}$$

$$\therefore \omega_{f2} = \frac{T_{d2} \times \omega_{d2}}{T_{f2}}$$

Substituting for ω_{d2} gives

$$\omega_{f2} = \frac{T_{d2} \times \omega_{d2}}{T_{f2}}$$

$$= \frac{T_{d1} \times \omega_{d1}}{T_{f1}} \times \frac{T_{d2}}{T_{f2}}$$

$$\therefore \frac{\omega_{f2}}{\omega_{d1}} = \frac{T_{d1} \times T_{d2}}{T_{f1} \times T_{f2}}$$

$$\therefore \frac{\omega_{d1}}{\omega_{f2}} = \frac{T_{f1} \times T_{f2}}{T_{d1} \times T_{d2}} = VR$$

$$VR = \frac{\text{product of all numbers of teeth on follower gears}}{\text{product of all numbers of teeth on driver gears}}$$

❏ *Example 5.15*

A gear system has a configuration similar to the compound gear train shown in Figure 5.12. The primary driver and the follower have 16 and 32 teeth, respectively. The secondary driver and follower have 16 and 30 teeth, respectively. Calculate the velocity ratio.

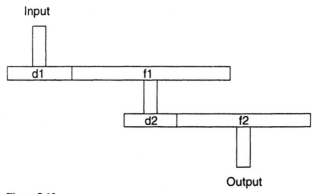

Figure 5.12

$$VR = \frac{\text{product of all the numbers of teeth on follower gears}}{\text{product of all the numbers of teeth on driver gears}}$$

$$= \frac{32 \times 30}{16 \times 16} = 3.75$$

Problems 5

1. A piston moves at a uniform velocity of 3.6 m/s against a resistance of 100 N. Find the power developed.
2. A vehicle of mass 1250 kg is raised through a height of 2 m in 30 s by a hydraulic lift. Calculate the work done.

3. A vehicle travels with a steady resistance to motion of 10 kN at a speed of 100 km/h. What power is required?

4. An engine develops a torque of 100 Nm at 2000 rev/min. What is the power of the engine?

5. A truck of mass 3500 kg climbs a 15° slope for 1 km. Calculate the work done in reaching the top of the slope.

6. A bike of mass 400 kg travels at a speed of 115 km/h. The bike then slows down until the speed is 75 km/h. Calculate the change in kinetic energy.

7. The mass of a flywheel is 17 kg and the radius of gyration is 0.122 m. Calculate the kinetic energy stored in it when the engine rotates at 900 rev/min.

8. A four cylinder, four stroke internal combustion engine has a diameter of 100 mm and a stroke length of 115 mm. The mean effective pressure is 750 kN/m^2 when the engine revolves at 2300 rev/min. Calculate the power produced.

9. A set of rope pulley blocks has three pulleys at the top and two at the bottom. An effort of 300 N is required to lift a load of 1.341 kN. Calculate the VR, the MA and the efficiency.

10. A Weston differential pulley block lifts a load of 500 N. The effort applied is 65 N and the efficiency is 45%. The diameter of the larger pulley is 150 mm. Calculate the diameter of the smaller pulley.

11. A compound gear train consists of a primary driver and follower having 15 and 40 teeth, respectively. The primary follower is keyed to the same shaft as the secondary driver. The secondary driver and follower have 17 and 52 teeth, respectively. Calculate the VR of the gear train.

12. A set of rope pulley blocks has four pulley wheels at both the top and bottom. When lifting a load of 14 kN the efficiency is 65%. Calculate the effort applied.

6 Thermodynamics

6.1 Introduction to thermodynamics

Thermodynamics is an engineering science that deals with energy conversion, particularly machines that convert heat into work. Thermodynamics is actually a general title and refers to a whole range of applications such as refrigeration, internal combustion engines and steam turbine power systems. An essential aspect of motor vehicle science is concerned with the efficient use of energy. We need to study this subject to understand the energy conversions that take place in a vehicle. For example, in an internal combustion cylinder, a fuel and air mixture is compressed and ignited. Very high pressures and temperatures are reached inside the cylinder. The pressure in the cylinder forces the piston down the cylinder and work is done by the engine mechanism. In this way chemical energy is converted into work.

This chapter looks at the basic principles of thermodynamics. The units involved have already been covered in previous chapters.

Thermodynamic systems

The term 'system' is commonly used throughout engineering and refers to something that is being investigated. A mechanical system is a body or a collection of components whose study involves motion and its causes (e.g. a pendulum moving with simple harmonic motion), or a state of equilibrium (e.g. forces bending a cantilever beam). A thermodynamic system is a region in space containing a quantity of matter whose behaviour is being investigated. For each problem it is necessary to define this region carefully, the system being separated from its surroundings by the **boundary**. The boundary may be fixed or elastic. Everything outside the boundary that may be affected by the system is known as the **surroundings**. There are two types of thermodynamic system: a **closed system** and an **open system**. With an open system, matter (e.g. gas) crosses the boundary as well as work and heat. With a closed system, the same matter remains within the boundary and only work and heat cross the boundary. Here are some examples of closed and open systems.

The inner surface of the fire extinguisher in Figure 6.1 is the boundary and its shape remains fixed. No mass crosses the system boundary (until it is used of course) and so the system is closed.

Figure 6.2 shows the piston and cylinder arrangement of a four stroke internal combustion engine. When all valves are closed, during the compression stroke, no mass crosses the system boundary and so the system is closed. As the piston moves up

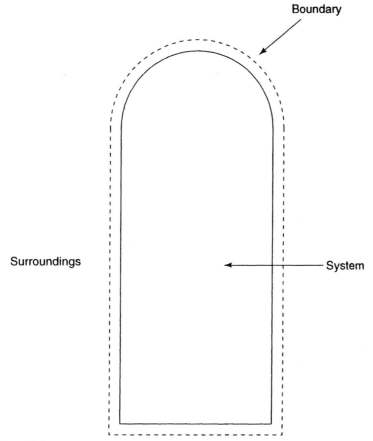

Figure 6.1

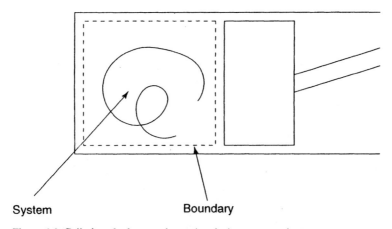

Figure 6.2 Cylinder of a four stroke engine during compression

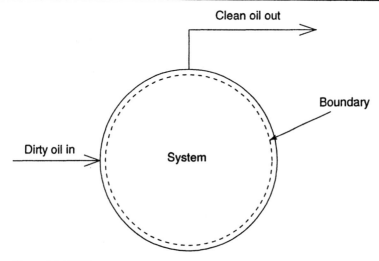

Figure 6.3 Oil filter

during the compression stroke, the boundary is elastic. This is only one of the four processes that make up the cycle of operation.

The oil filter in Figure 6.3 has a fixed boundary. Dirty oil passes into the system and clean oil passes out. As the oil flows through the system, the system is open.

The equations used in thermodynamics are based on the principles of conservation of mass, energy and momentum:

mass entering a system − mass leaving a system
= change of system mass
energy entering a system − energy leaving a system
= change of energy in system

This is the basis of the first law of thermodynamics and there will be more on this later.

For the principle of conservation of momentum we can apply Newton's second law of motion to an open system:

rate of change of momentum in − rate of change of momentum out
= external forces

In thermodynamics the term fluid is used to refer to liquids, vapours or gases.

Pressure, temperature, internal energy and the kinetic theory of gases

The kinetic theory of gases assumes that a gas is a large number of tiny hard spheres (molecules). These move about at high velocities at random. We can use this theory to explain pressure and temperature. Pressure is due to the force of the molecules hitting the container walls and is the average force per unit area on the container walls. Temperature is proportional to the average of the squares of the velocities of the molecules. This is also proportional to their kinetic energy. **Internal energy** is something you may not have come across before. Internal energy is the energy stored in a substance due to the motion of the molecules and their relative positions. It is the sum of the internal kinetic energies and potential energies of the molecules, and it can only increase or decrease if energy crosses the boundary, in or out of the system. This is a function of temperature. The symbol for internal energy is U and its units are the same as any energy, joules. There are other more complex theories that produce more information but the kinetic theory gives a good explanation of how gases behave.

Pressure

Thermodynamics usually deals with fluids and rarely solids. Pressure is the force that a fluid exerts on a specific area. It is defined as:

$$\text{pressure} = \frac{\text{force}}{\text{area}} \left[\frac{N}{m^2}\right]$$

The quantity symbol for pressure is P. The units of pressure are the same as those of stress in a solid, N/m^2. The pressure of a fluid is really the equivalent of stress of a solid. Other units you may come across are the pascal [Pa], and the bar [bar].

$1\,Pa = 1\,N/m^2$
$1\,bar = 10^5\,N/m^2$

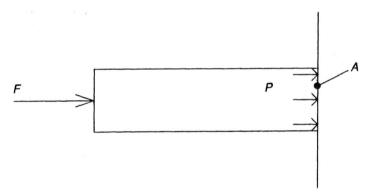

Figure 6.4

Figure 6.4 shows a bar pressed against a fixed wall by a force, F. If the area of contact is A then the pressure P at the contact surface is F/A. This pressure acts normal to the contact surfact. Figure 6.5 shows a piston and cylinder containing a fluid. As a force F is applied to the piston, which has cross-sectional area A, then the pressure in the fluid is F/A. The pressure now acts in all directions at right angles to the retaining walls.

The bar or millibar is often used in weather forecasts. One bar is approximately the pressure created by the atmosphere.

The only tricky part of pressure calculations, is the difference between **gauge pressure** and **absolute pressure**. Absolute pressure (P_{abs}) is measured as you would expect with a datum of $0\,N/m^2$. Many pressure gauges do not however measure absolute pressure, but measure using atmospheric pressure (P_{atm}) as the datum; this

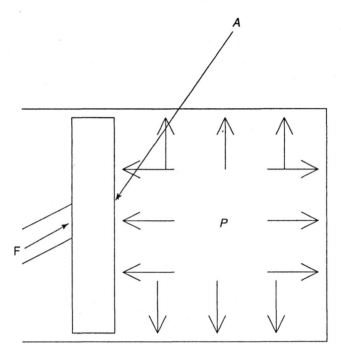

Figure 6.5

is known as gauge pressure (P_{gauge}). To convert gauge pressure into absolute pressure you need to add to it atmospheric pressure:

$$P_{\text{abs}} = P_{\text{gauge}} + P_{\text{atm}}$$

Atmospheric pressure varies slightly with time and location (see the weather forecast), and so for accurate calculations involving absolute pressure the latter must also be measured at the same time as the gauge pressure. The pressure of fluids and atmospheric pressure are sometimes expressed as a column of liquid, usually mercury or water.

Figure 6.6 represents a barometer and measures atmospheric pressure. The top end of the closed tube has a vacuum. Atmospheric pressure acts on the liquid at the base of the tube and liquid is pushed up the tube. As this is a system in equilibrium, the height of this liquid is directly proportional to the pressure at the base of the column, in this case atmospheric pressure. The relationship is as follows:

$$P = \rho g h \left[\frac{\text{kg}}{\text{m}^3}\right]\left[\frac{\text{m}}{\text{s}^2}\right][\text{m}] = \left[\frac{\text{N}}{\text{m}^2}\right]$$

where ρ is the density of the liquid used for the barometer, g is the acceleration due to gravity and h is the height of the column of liquid (see Section 6.5). If the top of the tube were broken and the vacuum lost then the height of the liquid would be the same as the level in the base, since the same pressure would act on both surfaces.

The U-tube in Figure 6.7 is called a manometer. It works in a similar way to the barometer, but instead of atmospheric pressure being applied to the tube, the vessel containing the fluid to be measured is applied; and at the top of the tube instead of

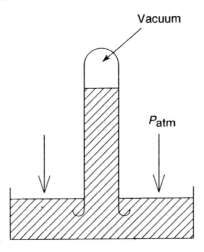

Figure 6.6

there being a vacuum, the tube is open to the atmosphere. At the liquid surface level in the left side of the U, the pressure will equal the pressure in the vessel. As the system is in equilibrium, the pressure at the same level in the right side of the U must be the same. This will equal atmospheric pressure plus the pressure due to the column of liquid. Therefore the manometer measures the pressure of the fluid above atmospheric pressure, i.e. gauge pressure. The same formula applies (ρgh). The important point to note is that the U-tube measures the difference in pressure between the two ends and this difference is represented by the height of the liquid.

If the U-tube manometer were used to measure pressures below atmospheric pressure, i.e. a vacuum, the liquid levels would look like those in Figure 6.8.

Atmospheric pressure acts at the liquid surface on the right side of the U. As the system is in equilibrium the pressure must be the same at the same level in the left

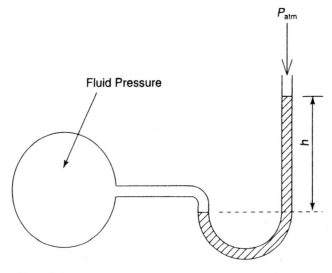

Figure 6.7

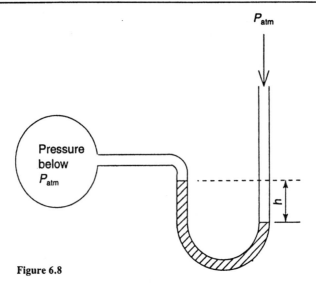

P_{atm}

Pressure
below
P_{atm}

h

Figure 6.8

side of the U. This will equal the pressure in the vessel plus the pressure due to the column of liquid. Therefore the height of the column of liquid represents the pressure below atmospheric pressure. In this case,

$$P_{abs} = P_{atm} - P_{gauge}$$

❏ Example 6.1

The gauge pressure of a gas in a vessel is measured as 5 m of water. Atmospheric pressure is measured at this time as 760 mm of mercury. Calculate the absolute pressure of gas. The relative density of mercury is 13.6.

$$P_{atm} = 13.6 \times 1000 \left[\frac{kg}{m^3}\right] \times 9.81 \left[\frac{m}{s^2}\right] \times 0.76 \, [m]$$

$$= 101\,396 \left[\frac{N}{m^2}\right] = 101.40 \left[\frac{kN}{m^2}\right]$$

$$P_{gauge} = 1000 \left[\frac{kg}{m^3}\right] \times 9.81 \left[\frac{m}{s^2}\right] \times 5 \, [m]$$

$$= 49\,050 \left[\frac{N}{m^2}\right] = 49.050 \left[\frac{kN}{m^2}\right]$$

$$P_{abs} = 49.05 \left[\frac{kN}{m^2}\right] + 101.40 \left[\frac{kN}{m^2}\right] = \mathbf{150.45 \; kN/m^2}$$

❏ Example 6.2

In a steam condenser there is a partial vacuum. The gauge pressure is measured as 709 mm of mercury. The barometer reading is 761 mm of mercury. Calculate the absolute pressure in the condenser. The density of the mercury is 13 600 kg/m³.

$$P_{abs} = P_{atm} - P_{gauge}$$
$$= (\rho g h)_{at} - (\rho g h)_g$$

As the same liquid is used for each gauge and the value of g is the same for each reading, the equation becomes:

$$P_{abs} = \rho g(h_{at} - h_g)$$

$$= 13\,600 \left[\frac{kg}{m^3}\right] \times 9.81 \left[\frac{m}{s^2}\right] \times (761 - 709) \, [mm] \times \frac{1}{1000}\left[\frac{m}{mm}\right]$$

$$= 6937.6 \text{ N/m}^2$$

Temperature

Temperature is a measure of the amount of heat transfer that will take place between a system and its surroundings. It is important that you do not confuse temperature with heat. Heat is the process of transferring energy across the boundary. This will occur when there is a temperature gradient present.

You are probably familiar with the temperature scales of Fahrenheit and Celsius or centigrade. A more useful scale in science and engineering is the **absolute temperature scale**. Think back to the kinetic theory of gases and temperature. If the temperature drops, the movement of the molecules slows down and the kinetic energy of the molecules will reduce. If the temperature continues to be reduced there will come a point where the molecules come to a standstill and have no kinetic energy. This temperature is known as absolute zero and is the coldest temperature possible. At this temperature the molecules will no longer hit the walls of the container and so the absolute pressure will also be zero. On the Celsius scale the temperature of absolute zero is −273.15°C. The unit of the absolute temperature scale is the kelvin and the unit symbol is K. Absolute zero is 0 K. An increment of 1 K is equal to 1°C. This means that a temperature on the Celsius scale can be converted to the kelvin scale by adding 273.15. For most problems, adding 273 will be accurate enough. For example convert 21°C to kelvin,

$$T(K) = T(°C) + 273 = 21°C + 273 = 294 \text{ K}$$

Also a temperature difference in the Celsius scale is the same as a temperature difference in kelvin. For instance a temperature gradient across a system boundary of 10°C is the same as a temperature gradient of 10 K.

❑ Example 6.3

A fluid enters a heater at 40°C and leaves at 93°C. Calculate the inlet and outlet temperature in kelvin, and the temperature difference in degrees Celsius and kelvin.

$$T(K) = T(°C) + 273$$
$$40°C + 273 = \textbf{313 K}$$
$$93°C + 273 = \textbf{366 K}$$
$$\text{temperature difference} = 93°C - 40°C$$
$$= \textbf{53 K}$$
$$\text{temperature difference} = 366 \text{ K} - 313 \text{ K}$$
$$= \textbf{53 K}$$

Thermodynamic properties and processes

A property is any characteristic of a system such as pressure, temperature, and internal energy. Some properties such as pressure and temperature are independent of the mass of the fluid within the system. A cup of coffee at 95°C has the same temperature as a car engine cooling system at 95°C. Properties independent of mass are called **intensive properties**. Properties that are dependent upon mass are called **extensive properties**. Examples of extensive properties are volume and internal energy. The volume of 2 kg of a fluid will be twice that of 1 kg of the same fluid under the same conditions. This can sometimes make calculations a little tricky. For this reason some properties are stated in **specific terms**. This means that the value of the property is related to a unit of mass. In the metric system, specific units are per kilogram. For example, if the volume of a gas is 2.61 m³ and the mass of this gas is 3 kg then the specific volume is 2.61/3 [m³/kg] = 0.87 m³/kg.

When a fluid undergoes a process, the change in the values of a property is a function of its initial and final values and has nothing to do with the process itself. For example, if a fluid undergoes a process and its temperature is raised from 300 K to 360 K, we know that the change in temperature is 60 K. We do not need to know about the process that caused the change, just the initial and final values.

When analysing thermodynamic systems, we assume that the fluid is in thermodynamic equilibrium. This means that the fluid is stable and the way changes occur is predictable. You will be familiar with a mechanical system in equilibrium, when all forces on a body balance. It is similar for a thermodynamic system. There are three requirements for the fluid to be in equilibrium:

1. The fluid must have internal thermal equilibrium. Although the fluid may undergo temperature changes as part of the process, the temperature throughout the fluid must be constant. If different temperatures exist within the fluid at the same time then this will cause heat transfer within the fluid and fluid movement within the boundary.
2. The fluid must be in chemical equilibrium. If the fluid changes chemically, it would be very difficult to analyse a situation.
3. The fluid must have pressure equilibrium. Different pressures within the fluid at the same time would cause unbalanced forces and a lack of mechanical equilibrium.

A thermodynamic process is the process through which fluid moves from one equilibrium state to another.

If a fluid goes through a series of processes and returns to its initial state, it has gone through a **thermodynamic cycle**. There are many ideal types of cycles used for analysing industrial processes and engines. These are usually divided into gas cycles or vapour cycles, depending upon the main working fluid. The internal combustion engine is usually defined as a gas cycle, since the main working fluid is combustion air. An example of a vapour cycle is a boiler steam cycle, which is described shortly.

Reversible processes

A fluid undergoes a process. If that fluid can then be taken back through all the stages of the process in a reverse order to reach the original state of the system and surroundings then the process is said to be **reversible**. This never happens in practice but it is often useful to assume some processes are reversible to simplify problems. Processes are always **irreversible** because of things like friction or churning within the system. There is a precise set of conditions that must be fulfilled to achieve a reversible system and this is always impossible.

Heat and work

Work is a form of energy transfer that takes place between a system and the surroundings due to the movement of a boundary (e.g. piston movement) or the rotation of a shaft. You will be familiar with the symbol for work, W. As it is an energy transfer process, its unit is the joule (J). In specific terms (i.e. for a unit of mass), the symbol is w and the units are J/kg.

Heat is also a form of energy transfer that takes place between the system and surroundings due to a temperature difference. Do not get confused between heat and temperature. The symbol for heat is Q and it is measured in joules, (J). In specific terms, the symbol is q and the unit is J/kg.

It is important to note that heat and work are not energies but are **energy transfer processes**. A system cannot possess heat or work. Heat or work can occur during a process. The usual terminology is heat transfer and work transfer. Heat and work are not properties either. A property depends only on initial and final values and is independent of the type of process. However, the magnitudes of heat and work do depend on the type of process the system undergoes.

The first law of thermodynamics

We will look initially at the first law of thermodynamics applied to closed systems. It will be applied to open systems in Section 6.4. Consider a steam power plant where a steam turbine drives a generator, as shown in Figure 6.9.

The whole plant is the process and so it is a closed cycle. Do not get confused by the fact that fluid is passing from one part of the process to another. No matter crosses the boundary of the process. Heat is applied to water in a boiler to produce steam and

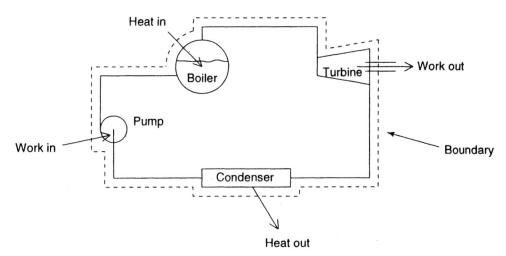

Figure 6.9 Steam power plant

so heat is transferred from the surroundings to the system. The steam is then used to drive a turbine and work is transferred to the surroundings. The steam is then condensed back to water and heat is transferred to the surroundings. The water is then pumped back into the boiler and work is transferred from the surroundings to the system. There are two main differences, as far as thermodynamics are concerned, between the cycles of an internal combustion engine and a steam power plant such as this one:

1. For an internal combustion engine the heat is supplied to the working fluid (air) inside the engine, whereas in the steam power plant the heat is supplied to the working fluid (steam) outside of the engine (the turbine).
2. The internal combustion engine is an open system since the working fluid flows through the system, whereas the steam power plant is a closed system and the working fluid is reused.

For each process that makes up the cycle, heat or work transfer may occur. As the fluid finishes the cycle in the same state that it started, there is no change of energy within the system. For a complete cycle of this process, the net heat transfer to the system is equal to the net work transfer to the surroundings.

This is the first law of thermodynamics when applied to a cycle. When a closed system undergoes a complete cycle, the algebraic sum of the work transfer is equal to the algebraic sum of the heat transfer, $Q_{net} = W_{net}$. The term net means that all the individual heat or work transfers for each process are added up. This is sometimes written as $(\Sigma W)_{cycle} = (\Sigma Q)_{cycle}$. The symbol Σ means that all the individual values are added up. At this point we need to establish some conventions for heat and work. These are the standard sign conventions used in thermodynamics:

- for heat transfer to the system, Q is positive
- for heat transfer from the system, Q is negative
- for work transfer to the system, W is negative
- for work transfer from the system, W is positive.

The above form of the first law of thermodynamics applies only to cycles. If a system undergoes a change of state that is not a cycle then the summation of energy transfers across a boundary is not necessarily zero. Go back to the principle of conservation of energy:

energy entering a system − energy leaving a system = change of energy in system

If we denote the change of system energy as ΔE, then the above statement can be written:

$$Q - W = \Delta E$$

ΔE can be assumed to be kinetic energy, potential energy and internal energy. Also, in a closed system the changes in kinetic and potential energies can be assumed to be negligible. In this case, the above equation becomes:

$$Q - W = \Delta U$$

When ΔU is positive the system gains internal energy; when it is negative the system loses internal energy. The usual form of the equation is:

$$Q = W + \Delta U$$

This is known as the non-flow energy equation (because it only applies to closed systems) and is true whether the process is reversible or not.

☐ Example 6.4

800 J of heat is transferred to the gas in a piston and cylinder system. If the gain in internal energy of the gas is 660 J, is the process an expansion or a compression? If the piston stroke is 0.2 m, what is the resistive force on the piston?

As the heat is transferred to the gas then Q is positive. As the change in internal energy is a gain then ΔU is positive.

$$W = Q - \Delta U$$
$$= 800 \, [J] - 660 \, [J]$$
$$= 140 \, J$$

As the answer is positive the work is transferred from the system to the surroundings and so the gas expands. The work done also equals the piston force multiplied by the stroke, l:

$$W = F \times l$$

$$F = \frac{W}{l}$$

$$= \frac{140 \, [J]}{0.2 \, [m]} = 700 \, N$$

Remember that $1 \, J = 1 \, Nm$.

☐ Example 6.5

A piston compresses air in a cylinder using 400 J of external work. The piston is insulated so that any heat transfer is negligible. What is the change of the internal energy?

Work is done on the system, so according to the convention the value is negative. The value of Q is zero.

$$\Delta U = Q - W$$
$$= 0 - (-400 \, [J]) = +400 \, J$$

The change of internal energy is an increase of 400 J.

6.2 Heat and temperature

When heat is transferred to mass, changes in its temperature and state can occur. It is important to know the relationships between heat, temperature and the state of different substances. In a motor vehicle, many temperature changes and heat transfers occur involving structural components, liquids and gases.

Specific heat capacity

Specific heat capacity is used to measure the amount of heat required to cause a change in temperature of a body. Heat that causes a change in temperature is called **sensible heat**. The specific heat capacity of a substance is the heat required to cause a unit temperature rise of a unit mass. In the metric system the units of specific heat capacity are J/kgK and its symbol is c. As this is concerned with the change in temperature then instead of the unit for kelvin, K, degrees Celsius could be used giving J/kg°C. This results in an identical calculation.

The formula we need to use is:

$$Q = m \, c \, \Delta\theta$$

Unit check: $[J] = [kg] \left[\dfrac{J}{kg \, K} \right] [K]$

where $\Delta\theta$ is the change in temperature

c is the specific heat capacity

m is the mass.

The actual value of c depends on the temperature at which it is measured. The variation is only slight, however, and it can be ignored for most calculations. The term specific means that it applies to a unit of mass. You may come across a heat capacity of a body rather than a specific heat capacity. This has the symbol C and the units are J/K. This means that the mass of the body has already been considered as part of the value. The formula to use then is:

$$Q = C\Delta\theta$$

Heat capacity is sometimes referred to as thermal capacity.

❑ Example 6.6

A piston has a mass of 0.5 kg. How much heat is required to raise the temperature of the piston from 15°C to 120°C? The piston consists of 0.2 kg of aluminium and 0.3 kg of steel.

The specific heat capacity of the steel, $C_{St} = 880 \left[\dfrac{J}{kg \, K} \right]$

The specific heat capacity of the aluminium, $C_{Al} = 510 \left[\dfrac{J}{kg \, K} \right]$

$$Q = mc\Delta\theta$$

$$Q_{total} = m_{St} \times c_{St} \times \Delta\theta + m_{Al} \times c_{Al} \times \Delta\theta$$

$$= \Delta\theta((m_{St} \times c_{St}) + (m_{Al} \times c_{Al}))$$

$$= (120 - 15) \, [K] \times \left[\left(0.2 \, [kg] \times 510 \left[\dfrac{J}{kg \, K} \right] \right) + \left(0.3 \, [kg] \times 880 \left[\dfrac{J}{kg \, K} \right] \right) \right]$$

$$= 105 \, [K] \times \left[102 \left[\dfrac{J}{K} \right] + 264 \left[\dfrac{J}{K} \right] \right]$$

$$= 38\,430 \, J = 38.4 \, kJ$$

Latent heat

As you will know, there are three forms a substance can take: solid, liquid and gas. These different forms are called phases. If heat is supplied to matter in the solid phases and the temperature increases, then eventually, the melting point will be reached. Further heat must be supplied to the solid for it to melt, during which time no further temperature rises occur.

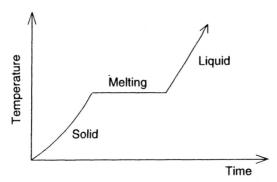

Figure 6.10 Temperature changes with time for a solid melting, with a steady heat input

The graph in Figure 6.10 shows temperature against time as a solid is heated through the melting point. The melting process is a flat horizontal line. This energy is used to provide the increased molecular energy that the liquid phase has and it is called **latent heat**. This is different from the heat used to cause a temperature rise. Once the solid has completely melted to a liquid then further additions of heat will cause an increase in the temperature of the liquid. The graph here is typical of a substance going through a phase change.

Latent heat is also required to convert a substance from the liquid phases into the gas phase. This change causes a large increase in volume and requires a much higher value of latent heat than the solid to liquid conversion. The molecules need sufficient energy to break free from the liquid phases and become a gas. The temperature at which this happens depends on the pressure on the surface of the liquid. Think of a pan of water on the stove. The molecules of a water surface are held back by the air pressure in contact with the surface. As the temperature rises and the water molecules try to break free they exert a pressure at that surface known as **vapour pressure**. When the vapour pressure overcomes the pressure of the air in contact with the surface, the ability of the air pressure to hold back the molecules of water reduces. If atmospheric pressure increases, the temperature at which the water boils increases as a higher vapour pressure is needed to allow molecules to escape. If atmospheric pressure reduces, so will the boiling temperature. Water boils at approximately 100°C on Earth at sea level because of the value of atmospheric pressure. Atmospheric pressure reduces with increased height above sea level, and so a lower vapour pressure is required for water to boil. Water therefore boils at a lower temperature up high mountains and climbers have difficulty making a pot of tea hot enough.

Engine cooling water can circulate at a temperature of 110°C without any problem if the system has sufficient pressure to increase the boiling temperature to above 110°C.

Even when a liquid is at a temperature that is much lower than boiling temperature, some vapour pressure is present and some molecules will escape causing evaporation. This is why clothes can dry out at relatively low temperatures on a washing line. The steam vapour that is produced from a liquid water surface such as a pan of water is actually a mixture of gas and water droplets. This is called **wet steam**. The mass of liquid present in the wet steam is measured with a **dryness fraction** from 0 to 1 so, for example, wet steam that is half comprised of water droplets would have a dryness fraction of 0.5. When the steam is a pure gas with no liquid present then the dryness fraction is 1. When water just reaches the boiling point and the dryness fraction is zero, this is known as a **saturated fluid**. Vapour that is completely dry with a dryness fraction of 1 is known as **saturated vapour**. There is more about this on page 222.

We need to be able to calculate the latent heat values. This can be done using the following:

Specific latent heat of fusion of a solid – The heat required to convert a unit of mass of the solid to liquid at the melting point without any temperature changes.
Specific latent heat of vaporisation of a liquid – The heat required to convert a unit of mass of the liquid to gas at the boiling point without any temperature changes. This value, like boiling temperature, depends on the pressure acting on the liquid surface.

The units of specific latent heat of vaporisation and fusion are kJ/kg.

❏ *Example 6.7*

1 kg of water is at atmospheric pressure and at boiling point, i.e. it is a saturated fluid. 2 MJ of heat is supplied. How much liquid is changed into a gas? If all the fluid is converted into wet steam what is the dryness fraction? The latent heat of vaporisation of water at this pressure is 2257 kJ/kg.

Let us represent the latent heat of vaporisation by the letter h. The heat required for the change of phase can be calculated from:

$$Q = h \times m$$

$$\therefore \text{ mass, } m = \frac{Q}{h}$$

$$= \frac{2 \times 10^3 \, [\text{kJ}]}{2257 \left[\dfrac{\text{kJ}}{\text{kg}}\right]} = 0.886 \text{ kg}$$

So 0.886 kg of the water is converted into steam. If all of the 1 kg of water becomes wet steam, 0.886 kg is steam and $(1 - 0.886) = 0.144$ kg is water droplets. The dryness fraction is therefore:

$$\frac{0.886 \, [\text{kg}]}{(0.886 + 0.114) \, [\text{kg}]} = 0.886$$

An example where the latent heat of phase change is put to good use is in a refrigerator. The refrigerator uses a vapour cycle and the commonest type of refrigerator uses vapour compression. The system comprises an evaporator, a

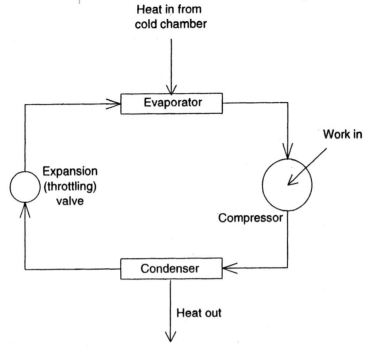

Figure 6.11 Refrigeration cycle

compressor, a condenser and an expansion valve (see Figure 6.11). The refrigerant enters an evaporator situated inside the refrigerator cabinet. At this point the refrigerant is a very wet vapour at a lower temperature than the inside of the refrigerator. Heat is taken from the refrigerator contents and the refrigerant leaves the evaporator as a much drier vapour. It then passes through a compressor which causes the pressure to rise and the temperature to rise above that of the surrounding room. The compressor also acts as a pump for the system. The heat from the refrigerant is passed to the surroundings by a condenser. The refrigerant is then changed to a wet vapour by an expansion valve and the cycle is repeated.

Heat transfer

So far we have talked about heat transfer without looking at how this happens. Heat transfer across a boundary can take place in three different ways: **convection**, **conduction** and **radiation**. All three are different but they all depend on a temperature difference for the transfer of energy.

Thermal conduction

Thermal conduction is heat transfer that takes place through a material. All materials can conduct heat in any state but it is only usually significant in solids. Energy is passed from one molecule of the material to the next. As an example of thermal conduction, think of touching the cylinder head of a running engine: it feels hot. This is because heat generated inside the engine is transferred to the surround-

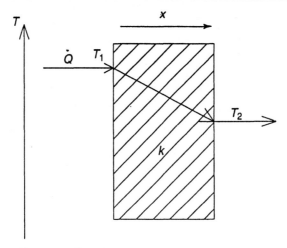

Figure 6.12 Single layer wall

ings by conduction through the cylinder head material. This is the easiest form of heat transfer to measure.

Usually heat transfer calculations involve **heat transfer rate** rather than just heat transfer because in engineering the rate of energy supply is important. In this case, the symbol for heat has a dot above it to show that time is involved. Heat transfer Q is expressed in joules (J). Heat transfer rate $\dot{Q}$ has units of joules per second (J/s) or watts (W). Consider heat flow through a flat wall (see Figure 6.12).

The temperature difference, ΔT ($= T_2 - T_1$), is the driving force of the heat transfer.

Heat transfer through some materials is easier than through others. How well a material conducts heat is a property called **thermal conductivity**. The symbol for thermal conductivity is k and the units are W/mK. Heat transfer rate also depends upon the thickness of the wall and the area through which the heat transfer takes place.

The basic thermal conduction equation, known as the Fourier equation, is as follows:

$$\dot{Q} = -kA\frac{\Delta T}{x}$$

A is the area through which the heat is transferred and x is the thickness of the wall.

Units check: $\left[\dfrac{J}{s}\right] = \left[\dfrac{W}{mK}\right] \times [m^2] \times \left[\dfrac{K}{m}\right] = [W]$

The minus sign is to keep to a convention. In equations, when the difference of two values is being calculated, the final value would be subtracted from the initial value. So, since the heat must flow from the higher temperature to the lower temperature across the wall, this would result in a minus sign for ΔT. This convention is maintained and the minus sign added to the equation to give a positive value of heat transfer rate.

Consider a wall made up of two layers of different materials of thermal conductivity, k_1 and k_2. If the temperature between the two layers is T, the temperature difference for the two materials is $(T_2 - T)$ and $(T - T_1)$. Heat that passes through one layer must also pass through the other layer, so the heat flow through each layer is the same:

$$\dot{Q} = -k_1 A \frac{(T - T_1)}{x_1} = -k_2 A \frac{(T_2 - T)}{x_2}$$

Rearranging these gives:

$$T = T_1 - \frac{\dot{Q}x_1}{Ak_1} \quad \text{and also} \quad T = T_2 + \frac{\dot{Q}x_2}{Ak_2}$$

$$\therefore T_1 - \frac{\dot{Q}x_1}{Ak_1} = T_2 + \frac{\dot{Q}x_2}{Ak_2}$$

$$\therefore (T_1 - T_2) = \frac{\dot{Q}x_2}{Ak_2} + \frac{\dot{Q}x_1}{Ak_1}$$

$$= \frac{\dot{Q}}{A}\left(\frac{x_2}{k_2} + \frac{x_1}{k_1}\right)$$

$$\therefore \dot{Q} = \frac{A(T_1 - T_2)}{\left(\dfrac{x_1}{k_1} + \dfrac{x_2}{k_2}\right)}$$

Notice that this has resulted in $(T_1 - T_2)$, i.e. the warmer temperature minus the cooler temperature, and that there is no minus sign at the front of the equation. The term x/k is called **thermal resistance**. The higher this value then the better the material is as an insulator. For any multi-layer wall consisting of n layers:

$$\dot{Q} = \frac{A(T_1 - T_n)}{\left(\dfrac{x_1}{k_1} + \dfrac{x_2}{k_2} + \cdots + \dfrac{x_n}{k_n}\right)}$$

where T_1 is the inner warmer temperature and T_n is the outer cooler temperature (see Figure 6.13). In some situations a high heat transfer rate is wanted; an example is

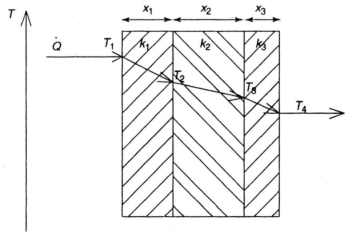

Figure 6.13 A wall consisting of three layers

through the walls of an engine cooling system radiator. For this, a material must be used that has a high thermal conductivity. If a material were used with a low thermal conductivity, the cooling system would not be very efficient. A material that is used because of its high thermal conductivity is called a **conductor**. Most metals are good conductors. A material that is a good electrical conductor is usually a good thermal conductor, e.g. copper and aluminium. Sometimes a material is needed with a low thermal conductivity. If you wanted to lag pipes to reduce their heat loss, a material that had a high thermal conductivity would not be much use. A material used because of its low thermal conductivity is called an **insulator**. Most non-metals are insulators. Particularly good insulators are asbestos, foam plastic and glassfibre.

❏ *Example 6.8*

A house wall consists of three layers. The outer layer is brick 100 mm thick with a thermal conductivity of 0.8 W/m K. The middle layer is cavity wall insulation 150 mm thick and has a thermal conductivity of 0.08 W/m K. The inner layer is breeze block 200 mm thick with a thermal conductivity of 0.6 W/m K. The temperature inside the house is 21°C. The temperature outside is 1°C. Calculate the heat loss through a wall of the house measuring 3 m by 4 m. Which layer is the best insulator?

$$\dot{Q} = \frac{A(T_1 - T_3)}{\left(\dfrac{x_1}{k_1} + \dfrac{x_2}{k_2} + \dfrac{x_3}{k_3}\right)}$$

$$= \frac{(3 \times 4)\,[\text{m}^2] \times (21 - 1)\,[\text{K}]}{\left(\dfrac{0.2}{0.6} + \dfrac{0.15}{0.08} + \dfrac{0.1}{0.8}\right)\left[\dfrac{[\text{m}]}{[\text{W/m K}]}\right]}$$

$$= \frac{12\,[\text{m}^2] \times 20\,[\text{K}]}{(0.333 + 1.875 + 0.125)\left[\dfrac{\text{m}^2\,\text{K}}{\text{W}}\right]}$$

$$= \mathbf{102.857\ W}$$

The cavity wall is the best insulator, as it has the highest value of thermal resistance: 1.875 m² K/W.

Thermal convection

Thermal convection is a form of heat transfer that occurs due to the movement of a fluid (gas or liquid). In an engine, heat is transferred from hot parts of the engine to the radiator by the circulation of the coolant. This is convection. Fluids usually move through pressure differences caused by a pump or because of density changes caused by temperature differences.

Consider an engine cooling system, as shown in Figure 6.14. The pump creates a pressure increase and forces the cooling water to circulate around the engine cylinders. The water is at a lower temperature than the engine and so heat is transferred from the engine to the water. The water then flows around the system to the radiator tubes. Here, air flowing around the tubes takes heat from the water. The cooling water then flows back to the pump for another lap. Heat is transferred (from

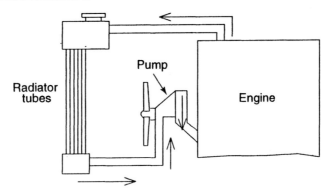

Figure 6.14 Engine cooling system

the engine to the radiator) because the pressure difference created by the pump causes the particles of fluid to move.

Consider now a water heater and radiator (Figure 6.15). When water undergoes a temperature increase, its density reduces (except at very low temperatures). The water leaves the heater at a higher temperature than the water in the rest of the system. It therefore has the lowest density and the water rises to the radiator. At the radiator, the water loses heat and the temperature of the water reduces. The density of the water then increases and it flows down back to the heater. Heat is transferred as particles of fluid move from the heater to the radiator due to the changes in density.

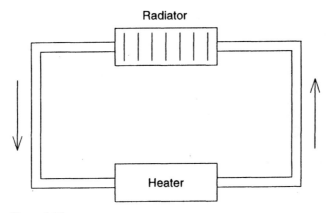

Figure 6.15

Thermal radiation

Thermal radiation can take place without a transfer material, unlike conduction and convection. Radiation is energy emitted by electrons vibrating at the surface of a body. A good example of this is heat from the Sun reaching the Earth. There are no materials between the Earth and the Sun but plenty of energy reaches us. Radiation can pass through matter (but not as well as through space); the extent of this depends upon the properties of the material. With thermal conduction, the amount of heat transferred depends upon the temperature difference. With thermal radiation, the heat transferred mainly depends upon the absolute temperature of the emitting surface. All materials emit or absorb radiation to some extent, provided that their temperature is above absolute zero. A body with a surface that absorbs all radiant heat it receives is called a black body. 'Black body' is a thermal name and has nothing to do with colour. As well as absorbing energy a black body also emits the maximum amount of radiation possible. A black body is an ideal theoretical concept and in reality, bodies always absorb or emit thermal radiation at a lower level than a black body. The level of energy absorbed or emitted relative to a black body is accounted for in calculations by an emissivity factor, which is between 0 and 1. A black body has an emissivity factor of 1. A rough surface has an emissivity factor of around 0.7–0.8, and a shiny surface has an emissivity factor of around 0.1.

Table 6.1 Typical emissivity values of some common materials

Material	Emissivity
Polished aluminium	0.06
Dull aluminium	0.1
Polished steel	0.35
Dull steel	0.6
White paint	0.8
Matt black paint	0.95

From Table 6.1 you will see that colour has less to do with emissivity than texture. A white car on a hot summer's day will not necessarily remain cooler than a black car. The car that has the shiniest smoothest paint surface will have the lowest emissivity factor and will absorb the least thermal radiation. Generally smooth shiny surfaces have low emissivity factors and rough shiny surfaces have high ones. Mirrors or polished copper have low emissivity and absorptivity factors and so are good reflectors of radiant heat.

6.3 Gases

With most subjects in science and engineering, we have to make various assumptions to simplify calculations and apply theory. In thermodynamics, we assume a fluid is in thermodynamic equilibrium or sometimes that a process is reversible. With gases, a useful concept is a **perfect gas**. A perfect gas is a simplified version of real gases to make calculations easier. Gases obey various laws and have properties that we shall be looking at. A perfect gas obeys these laws and has constant property values, such as specific heat capacities that do not vary with temperature. The assumptions made are:

1. the gas is homogeneous (this means it is the same throughout; it is perfectly mixed up)
2. the gas is made up of lots of very tiny rapidly moving molecules that are perfectly elastic (when hitting container walls or each other) and take up negligible space
3. the gas will spread to fill whatever container it is in.

For these assumptions to be made, the gas must be at a temperature well above the saturation temperature. At extreme pressures and temperatures the molecules themselves may take up too much room for assumption 2 to be applied.

Boyle's law and Charles' law

Boyle's law states that, if a mass of gas is kept at a constant temperature then the pressure is inversely proportional to the volume. Pressure is represented by p and volume by V. Mathematically then this is:

$$p \propto \frac{1}{V}$$

or $V \propto \dfrac{1}{p}$

or $p \times V = \text{constant}$

So if a gas following Boyle's law changes from pressure p_1 and volume V_1 to pressure p_2 and volume V_2 at a constant temperature then:

$$p_1 \times V_1 = p_2 \times V_2$$

☐ Example 6.9

A cylinder contains air at a pressure of 100 kN/m². The volume is 0.05 m³. The piston moves to reduce the volume to 0.025 m³ whilst the temperature is kept constant. What is the new pressure? Assume that the air behaves as a perfect gas.

As the temperature is kept constant during the operation, Boyle's law can be applied:

$$p_1 \times V_1 = p_2 \times V_2$$

$$\therefore p_2 = \frac{p_1 V_1}{V_2}$$

$$= \frac{100 \times 10^3 \left[\dfrac{N}{m^2}\right] \times 0.05\ [m^3]}{0.025\ [m^3]}$$

$$= 200 \times 10^3\ \text{N/m}^2 = \mathbf{200\ kN/m^2}$$

Charles' law states that if a mass of gas is kept at a constant pressure then the volume is directly proportional to the temperature. Mathematically this is:

$$V \propto T$$

or $T \propto V$

or $\dfrac{V}{T} = \text{constant}$

So if a gas following Charles' law changes from temperature T_1 and volume V_1 to temperature T_2 and volume V_2 at a constant pressure then:

$$\frac{V_1}{T_1} = \frac{V_2}{T_2}$$

Example 6.10

A balloon contains a gas at a pressure of 50 kN/m^2 and at a temperature of 20°C. The temperature is raised to 100°C whilst the pressure is kept constant. What happens to the volume? Assume that the gas behaves as a perfect gas.

As the pressure is kept constant then Charles' law may be applied.

$$\frac{V_1}{T_1} = \frac{V_2}{T_2}$$

$$\therefore V_2 = \frac{V_1 T_2}{T_1}$$

Unfortunately, we do not know the initial value of the volume.

$$V_2 = \frac{V_1 \times (100 + 273)\,[\text{K}]}{(20 + 273)\,[\text{K}]}$$

$$= V_1 \times 1.273$$

$$\therefore \frac{V_2}{V_1} = 1.273$$

We can say then that the volume increases by **1.273 times.**

Combining the laws

Boyle's law, $p \times V = $ constant at a constant temperature, and Charles' law, $V/T = $ constant at a constant pressure, can be combined to give the equation:

$$\frac{pV}{T} = \text{constant}$$

This means that for a fixed mass of gas the relationship between the gas pressure, volume and temperature remains the same. If a gas undergoes a process from condition 1 to condition 2, then the relationship between pressure, volume and temperature is:

$$\frac{p_1 V_1}{T_1} = \frac{p_2 V_2}{T_2}$$

Example 6.11

A gas is contained in a cylinder at a pressure of 50 kN/m^2, a temperature of 20°C and a volume of 8 × 10^{-3} m^3. The piston compresses the gas to a volume of 3 × 10^{-3} m^3. The new temperature is 100°C. Calculate the new pressure.

$$\frac{p_1 V_1}{T_1} = \frac{p_2 V_2}{T_2}$$

$$\therefore p_2 = \frac{p_1 V_1 T_2}{T_1 V_2}$$

$$= \frac{50 \times 10^3 \left[\frac{N}{m^2}\right] \times 8 \times 10^{-3} [m^3] \times (100 + 273) [K]}{(20 + 273) [K] \times 3 \times 10^{-3} [m^3]}$$

$$= 169\,738.34 \text{ N/m}^2$$

$$= \mathbf{169.7 \text{ kN/m}^2}$$

Characteristic gas equation

The example above deals with a fixed mass of gas. The value of the mass is not important for the calculation. You may come across a process that does not involve a fixed mass of gas. The formula above sometimes needs expanding to take into account a change of mass. If the combined law is applied to a mass of gas of 1 kg then the volume V has a particular value; this is the specific volume and is usually written as v. Remember that specific means per unit of mass. In this case the specific volume is the volume for one kilogram of mass and the units are kg/m^3 rather than just kg. The combined equation can be written for one kilogram of mass using the specific volume:

$$\frac{pv}{T} = \text{constant}$$

The constant in this case is special and is unique for each different gas. It is called the specific gas constant and given the symbol R.

$$\frac{pv}{T} = R$$

The units of R can be calculated as follows:

$$\left[\frac{N}{m^2}\right]\left[\frac{m^3}{kg}\right]\left[\frac{1}{K}\right] = \left[\frac{N\,m}{kg\,K}\right] = \left[\frac{J}{kg\,K}\right]$$

The combined equation can now be written as: $pv = RT$.

This still doesn't help us deal with different quantities of mass, but if we now multiply each side by the mass m:

$$p \times v \times m = m \times R \times T$$

But $v \times m = V$ so now:

$$p \times V = m \times R \times T$$

This equation, $pV = mRT$ is called the **characteristic gas equation** or the ideal gas equation of state. With this, the relationship between the gas pressure, volume and temperature of a process can be calculated when there is a change in the value of the mass.

❑ *Example 6.12*

A room measures 5 m by 3 m by 2.5 m. It contains air at a pressure of 1.01 bar. The temperature is 20°C. Calculate the mass of air in the room. Assume that the value of the specific gas constant R for air is 0.287 kJ/kg K.

Applying the characteristic gas equation,

$$p \times V = m \times R \times T$$

$$\therefore \ m = \frac{pV}{RT}$$

$$= \frac{1.01 \,[\text{bar}] \times 10^5 \left[\dfrac{\text{N}}{\text{m}^2\,\text{bar}}\right] \times (5 \times 3 \times 2.5) \,[\text{m}^3]}{287 \left[\dfrac{\text{J}}{\text{kg K}}\right] \times (20 + 273) \,[\text{K}]}$$

$$= \textbf{45.040 kg}$$

An introduction to thermodynamic property tables

The properties of steam and various other fluids are set out in thermodynamic property tables. A series of these is available in the form of a small booklet. These are commonly known as **steam tables**, although many other fluids are covered. Fluid data can be looked up in the tables relating absolute fluid pressure p_s and saturation temperature T_s to such properties as specific volume v and specific internal energy u. Other properties recorded in the tables are enthalpy h and entropy s, but although these are of great interest to engineers studying, for instance, steam power plant, they need not concern us. When fluid temperatures are way above the saturation temperature, the gas laws we have discussed may be used. When temperatures are too close to the vapour region though the gas laws do not apply, and the steam tables are then useful.

Look at the set of tables referring to saturated water and steam in Table 6.2. The far left column lists pressures starting from a very low pressure measured in bar to a very high pressure. Select a pressure, e.g. 1 bar, which is just below normal atmospheric pressure. Follow the line to the right: the next column records the saturation temperature T_s, which in this case is 99.6°C. This is the saturation temperature (boiling point) at that fluid pressure. Following the line further to the right: the next column records specific volume v_g, in this case 1.694 m³/kg. The

Table 6.2 Typical values from steam tables

p	t_s	v_g	u_f	u_g	h_f	h_{fg}	h_g
bar	°C	m³/kg	kJ/kg	kJ/kg	kJ/kg	kJ/kg	kJ/kg
1.0	99.63	1.694	417	2505	417	2258	2675
1.2	104.8	1.428	439	2512	439	2244	2683
1.4	109.4	1.236	458	2517	458	2232	2690
1.6	113.4	1.091	475	2521	475	2221	2696
⋮	⋮	⋮	⋮	⋮	⋮	⋮	⋮

subscript g means that this is the specific volume of the fluid when it is a dry saturated steam (just a gas with no wet steam) at that pressure. The next two columns are the internal energy of the saturated water u_f, and the internal energy of the saturated vapour u_g, at that pressure. The subscript f means a saturated liquid. It is important to remember the difference between a saturated liquid and a saturated vapour. A saturated liquid is liquid, in this case water, that has just reached its boiling point at the fluid pressure (this happens at about 100°C at normal atmospheric pressure). A saturated vapour is a completely dry vapour when all the liquid droplets have been converted into a gas, at the same temperature. In between is the progressive vapour stage, where a substance is midway between a liquid and a gas, with a dryness fraction between 0 and 1.

Other fluid tables usually included are ammonia, refrigerants and mercury.

❑ *Example 6.13*

What is the specific volume of 10 kg of steam when it is a dry saturated vapour at a pressure of 2 bar? What is the volume, the saturated temperature and the internal energy?

From the tables:

$$v_g = 0.8856 \left[\frac{m^3}{kg}\right]$$

$$V = mv_g = 10\,[kg] \times 0.8856 \left[\frac{m^3}{kg}\right]$$

$$= 8.856\ m^3$$

$$T_s = 120.2°C$$
$$= 120 + 273 = 393\ K$$

$$u_g = 2530 \left[\frac{kJ}{kg}\right]$$

$$U = mu_g$$

$$= 10\,[kg] \times 2530 \left[\frac{kJ}{kg}\right]$$

$$= 25\ 300\ kJ$$

Notice that u_g and v_g are used rather than u_f and v_f, as the steam is a dry saturated vapour and not a saturated fluid.

6.4 Steady flow processes

The changes that occur in the state of a thermodynamic system are called processes. Open and closed systems have already been discussed. When a process occurs in a closed system, energy may be transferred across the boundary as work or heat, but the working fluid never crosses the boundary. The processes that can occur are called non-flow processes and the non-flow energy equation may be applied:

$$Q = W + \Delta U$$

In an open system, the working fluid can also cross the boundary as well as energy. This means that the equation above does not apply, as other energy may cross the boundary with the working fluid. A process that occurs in an open system can be classed as either a steady flow process or an unsteady flow process (which are self-explanatory). We will assume that the flow processes here are steady. Many common flow processes can be investigated by assuming a steady flow. A fan heater is an example of a steady flow process, the working fluid being air steadily flowing in and out of the system. The conditions necessary to make this steady flow assumption are as follows:

1. the mass of working fluid flowing past any section of the system must have a constant flow rate
2. the properties of the fluid at any section of the system must remain constant during the process
3. all heat and work transfers must take place at a constant rate.

Many processes are flow processes: in a vehicle there are several systems all with a working fluid being pumped around. Consider the flow of exhaust gases through a turbocharger: the mass flow of exhaust gases into the turbine is equal to the mass flow out. The hot gases drive a gas turbine and cause work to be done on a drive shaft. The shaft causes work to be done on another flow process, the air compressor that supercharges the engine. Compressed air leaves the outlet of the compressor at the same mass flow rate as it enters the inlet. To analyse the process we need to select operating conditions when we can assume that the flow through the compressor is steady and does not vary. This highlights two steady flow processes. A gas turbine extracts work from a gas as it expands and reduces in pressure. A compressor is considered to be a turbine operating in reverse. Work is done on a fluid as it is compressed from a lower pressure to a higher pressure (Figure 6.16).

Another example of a steady flow process is a heat exchanger. This is a device, as the title suggests, for exchanging heat from one fluid to another. The two fluids are kept separate and the heat transfer takes place through the material walls. A car radiator is a heat exchanger. Heat is transferred from the jacket cooling water of the engine to a flow of air. A boiler is a type of heat exchanger: heat is transferred from hot combustion gases in the combustion chamber to water. Again note that the two fluids are kept separate.

The study of steady flow processes is based on the principle that energy cannot be created or destroyed, but only converted from one form to another. The equation used for analysing steady flow processes is known as the **steady flow energy equation** and is derived from an energy balance. An energy balance of a system implies that the total amount of energy entering a system per second is equal to the total energy leaving a system per second. Figure 6.17 represents an open system.

The working fluid enters the system through the inlet pipe at a constant rate. Inside the system various energy transfers can take place. For instance, for a boiler, heat would cross the boundary and enter the system; for an engine, work would cross the boundary and leave the system. When all energy transfers have taken place, the fluid

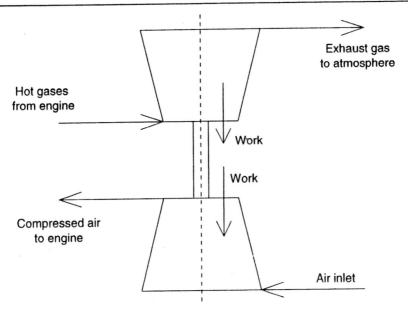

Figure 6.16

will leave the system through the outlet pipe. To calculate an energy balance, we need to consider all the energy crossing the boundary in a unit of time. Energy entering the system will be made up of:

- any heat transfer crossing the boundary per second
- the energy of the fluid itself flowing into the system per second; this will include internal energy, kinetic energy and potential energy.

If the mass flow rate is $\dot{m}$ kg/s then, at the inlet to the system,

$$\text{energy of the fluid} = \text{internal energy} + \text{potential energy} + \text{kinetic energy}$$
$$= \dot{m}u + \dot{m}gz_1 + \tfrac{1}{2}\dot{m}c_1^2$$

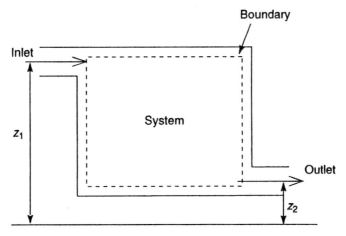

Figure 6.17

By using specific terms throughout and by multiplying all terms by the mass flow rate, we find the energy flow rate. The first term $\dot{m}u$ is the specific internal energy (i.e. per kilogram) multiplied by the mass flow rate to give the rate of internal energy entering the system. The second term $\dot{m}gz_1$ is the equation for potential energy using $\dot{m}$ instead of just m, and z instead of h for reasons that will become apparent. The third term $\frac{1}{2}\dot{m}c_1^2$ is the rate of kinetic energy entering the system using c for the velocity instead of v, as we will be dealing with specific volume and the two could become confused.

The fluid entering the system is pushed by the fluid behind it. This means that work is done on the fluid in the system. Fluid that leaves the system is pushed by fluid behind it. Therefore work leaves the system. The work done in moving the fluid is called the **flow work** or **pressure energy**. The work done by one kilogram of fluid is as follows:

work done = force × distance

Force exerted by the fluid = pressure × area = $p \times A$ (as $p = F/A$)

So, work done = $p \times A \times S$

where S is the distance moved to push one kilogram of the fluid.

Now, as $A \times S$ contains one kilogram of the fluid, this is the specific volume v.

work done = $p \times v$

The total energy entering the system needs to include net heat transfer and net work transfer as well as the energy of the fluid. These can now be added to the equation as follows:

total energy entering a system per second = $\dot{Q} + \dot{m}\left(u_1 + gz_1 + \frac{1}{2}c_1^2 + p_1v_1\right)\left[\frac{J}{s}\right]$

In a similar way, the total energy leaving the system per second will consist of the energy of the fluid leaving the system per second and any work leaving the system per second. The net work leaving the system per second across the boundary, is represented by $\dot{W}$ (not to be confused with flow work pv leaving the system, which is part of the energy of the fluid). The heat transfer is not considered since the heat transfer rate $\dot{Q}$ is the net heat transfer rate, i.e. the difference between heat in and heat out. Notice the same convention applies: heat transfer to the system is positive and work transfer to the surroundings is positive.

Total energy leaving the system per second = $\dot{m}\left(u_2 + gz_2 + \frac{1}{2}c_2^2 + p_2v_2\right) + \dot{W}\left[\frac{J}{s}\right]$

The principle of conservation of energy states that the total energy entering a system is equal to the total energy leaving a system. The two energy flow equations can now be combined in the same equation.

$$\dot{Q} + \dot{m}\left(u_1 + gz_1 + \frac{1}{2}c_1^2 + p_1v_1\right)\left[\frac{J}{s}\right] = \dot{m}\left(u_2 + gz_2 + \frac{1}{2}c_2^2 + p_2v_2\right) + \dot{W}\left[\frac{J}{s}\right]$$

This is normally written as:

$$\dot{Q} - \dot{W} = \dot{m}((p_2v_2 - p_1v_1) + (u_2 - u_1) + \tfrac{1}{2}(c_2^2 - c_1^2) + g(z_2 - z_1))\left[\frac{J}{s}\right]$$

This is the full form of the steady flow energy equation, SFEE. This looks like a large complicated equation but in many situations some of the terms can be assumed to be zero and so can be ignored, as you will see. If you work out the units of any term in the equation you will get the units of energy flow, joule per second (J/s):

$$\text{e.g.} \quad pv\dot{m} \Rightarrow \left[\frac{N}{m^2}\right]\left[\frac{m^3}{kg}\right]\left[\frac{kg}{s}\right] = \left[\frac{Nm}{s}\right] = \left[\frac{J}{s}\right]$$

The term $pv + u$ has a special meaning and name. It is called specific enthalpy and given the symbol h. We will not go into any detail on enthalpy in this book but it is important that you recognise the symbol and what it means because data for problems is sometime given in this form. So $h = pv + u$ and the units are the same as those for any specific energy terms (J/kg) or perhaps (kJ/kg). This is why z is used instead of h for height, to avoid confusion with enthalpy. The SFEE can be rewritten using h_1 instead of $p_1v_1 + u_1$ and h_2 instead of $p_2v_2 + u_2$:

$$\dot{Q} - \dot{W} = \dot{m}((h_2 - h_1) + \tfrac{1}{2}(c_2^2 - c_1^2) + g(z_2 - z_1))\left[\frac{J}{s}\right]$$

The total enthalpy of a system is sometimes stated in the terms below rather than the specific form: $H = pV + U$ [kJ].

☐ Example 6.14

The cooling water of a vehicle engine passes through a heater. The enthalpy of the water reduces by 20.1 kJ/kg across the heater. The water mass flow rate is 2.86 kg/min. Calculate the power of the heater.

We can assume that the kinetic energy of the water is the same at the outlet of the heater to the inlet, because the pipe diameter will be approximately the same. We can assume that the potential energy will be the same too, since the inlet and the outlet of the heat exchanger must be at a similar height. As there is no movement of the boundary or a rotation of a shaft then there will be no work transfer.

$$\dot{Q} - \dot{W} = \dot{m}((h_2 - h_1) + \tfrac{1}{2}(c_2^2 - c_1^2) + g(z_2 - z_1))\left[\frac{J}{s}\right]$$

For this problem the equation becomes:

$$\dot{Q} = \dot{m}(h_2 - h_1)$$

As the enthalpy reduces then $(h_2 - h_1)$ is negative:

$$\dot{Q} = 2.86\left[\frac{kg}{min}\right] \times \frac{1}{60}\left[\frac{min}{s}\right] \times -(20.1 \times 10^3)\left[\frac{J}{kg}\right]$$

$$= -958.1 \text{ W}$$

The negative sign indicates heat transfer from the system to the surroundings.

☐ Example 6.15

A hot air blower consists of an electrical heating element and a fan. The power input to the heating element is 1400 W. The gain in enthalpy of the air through the heater is

150 kJ/kg. The velocity of the air from the blower is 25 m/s. The mass flow through the heater is 10×10^{-3} kg/s. Calculate the power input to the fan.

$$\dot{Q} - \dot{W} = \dot{m}[(h_2 - h_1) + \tfrac{1}{2}(c_2^2 - c_1^2) + g(z_2 - z_1)]$$

Assume that the kinetic energy of the air at the inlet is zero and that there is no change in potential energy of the air through the system.
The equation becomes:

$$\dot{Q} - \dot{W} = \dot{m}[(h_2 - h_1) + \tfrac{1}{2}c_2^2]$$

The rate of work to the fan, i.e. the power input, is negative as it is transferred to the system:

$$-\dot{W} = \dot{m}((h_2 - h_1) + \tfrac{1}{2}c_2^2) - \dot{Q}$$

$$= 10 \times 10^{-3} \left[\frac{kg}{s}\right]\left(150 \times 10^3 \left[\frac{J}{kg}\right] + \frac{25^2}{2}\left[\frac{J}{kg}\right]\right) - 1400\left[\frac{J}{s}\right]$$

$$= 1.503 \times 10^3 \left[\frac{J}{s}\right] - 1400\left[\frac{J}{s}\right] = 103.1 \text{ J/s} = \mathbf{103 \text{ W}}$$

☐ *Example 6.16*

A fluid flows through a steady flow system at a rate of 0.5 kg/s. The initial pressure, volume and velocity are 15 bar, 150×10^{-3} m³/kg and 75 m/s respectively. The final pressure, volume and velocity are 1 bar, 1 m³/kg and 125 m/s respectively. The system transfers 200 MJ/h of heat to the surroundings. There is no work transfer and you can assume that the potential energy remains constant through the process. Calculate the change in internal energy.

$$\dot{Q} - \dot{W} = \dot{m}\{(h_2 - h_1) + \tfrac{1}{2}(c_2^2 - c_1^2) + g(z_2 - z_1)\}$$

The equation becomes:

$$\dot{Q} = \dot{m}\{(p_2v_2 - p_1v_1) + (u_2 - u_1) + \tfrac{1}{2}(c_2^2 - c_1^2)\}$$

Rearrange the equation to find the internal energy:

$$(u_2 - u_1) = \frac{\dot{Q}}{\dot{m}} - (p_2v_2 - p_1v_1) - \tfrac{1}{2}(c_2^2 - c_1^2)$$

$$= \left\{\frac{-200 \times 10^6 \left[\frac{J}{h}\right] \times \frac{1}{3600}\left[\frac{h}{s}\right]}{0.5\left[\frac{kg}{s}\right]}\right\}$$

$$- \left\{1 \times 10^5 \left[\frac{N}{m^2}\right] \times 1\left[\frac{m^3}{kg}\right] - 15 \times 10^5 \left[\frac{N}{m^2}\right] \times 150 \times 10^{-3}\left[\frac{m^3}{kg}\right]\right\}$$

$$- \frac{1}{2}\left\{125^2 \left[\frac{m}{s}\right]^2 - 75^2 \left[\frac{m}{s}\right]^2\right\}$$

$$= (-111.111 + 125 - 5) \times 10^3 \left[\frac{J}{kg}\right] = \mathbf{8.889 \times 10^3 \text{ J/kg}}$$

Notice that this is specific internal energy and the units are per kilogram of the working fluid.

The boundary of a process must be chosen carefully in order to apply the SFEE. For instance, a cylinder of a four-stroke internal combustion engine cannot be considered a steady flow process. See Figure 6.18. The fuel mixture enters the cylinder from the carburettor or the fuel injection system when the inlet valves are open. The valve then closes and the fuel mixture is compressed and ignited to cause the force on the piston. On the final stroke, the exhaust valve opens and the exhaust gases are forced out, before the process is repeated. This makes the mass flow of the process far from steady. By considering the boundary to be further outside the engine, the SFEE can be applied.

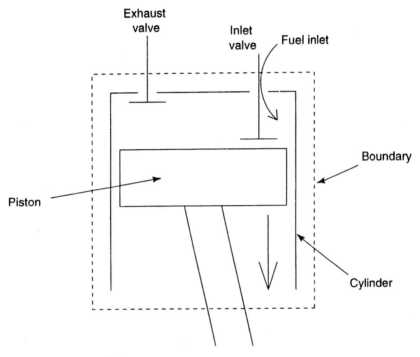

Figure 6.18 Non-steady flow process

Now the mass flow through the engine can be considered to be steady and the SFEE can give answers about such things as fuel consumption and work done by the engine. As the system boundary is imaginary anyway, then it can be moved to any position to simplify a system and we can assume a steady flow of mass and energy across the system boundary (Figure 6.19). The car as a whole could be considered to be a steady flow process. We have to adapt the steady flow energy equation slightly. There is no precise way of using the equation. The important thing to remember is to consider the energy entering or leaving the system per second.

When placing the boundary around the entire car like this we can ignore details such as the combustion process in the engine as these will all take place inside the car boundary. The fluid entering the system is the fuel/air mixture, and this leaves as exhaust gases. The problem can be greatly simplified by considering the dominant

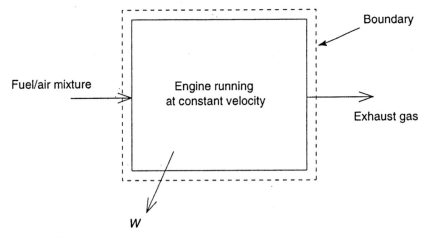

Figure 6.19

energy transfers taking place across the boundary. The energy flowing into the system can be considered to be due to the energy of the fuel, i.e. the petrol or diesel oil. All other energy entering the system, from kinetic energy of the fuel for example, can be assumed to be negligible. The energy leaving the system will be as heat transfer to the surroundings from the engine, work transfer to the surroundings as the car moves, and the energy of the fluid (i.e. exhaust gas) leaving the system. The majority of the energy leaving the system (if the process is efficient) will be due to the work that the car does (Figure 6.20).

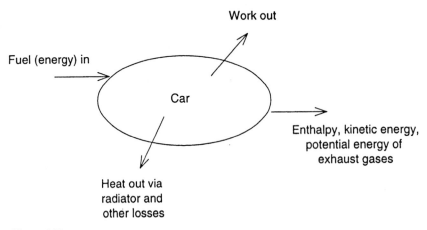

Figure 6.20

The total energy of the fuel can be measured by the **calorific value**. This is the amount of energy that any fuel can provide. It is measured as energy per unit mass, e.g. [MJ/kg].

☐ *Example 6.17*

A car fuel tank holds 50 kg of fuel with a calorific value of 45 MJ/kg. What is the total energy available from the fuel for the engine?

Energy available $= 50\,[\text{kg}] \times 45\,[\text{MJ/kg}]$
$$= 2250\,\text{MJ} = 2.25\,\text{GJ}$$

☐ *Example 6.18*

A car with a mass of 1400 kg travels along a flat road at a constant velocity and uses fuel of calorific value 42 MJ/kg at a rate of 6.12 kg/h. The car reaches a hill and climbs at a vertical rate of 0.5 m/s. Calculate the increase in fuel as a percentage. Assume that any enthalpy, kinetic energy and potential energy out of the car due to exhaust gases are negligible.

Consider the car on the flat road:

$$\dot{Q} - \dot{W} = \dot{m}[(h_2 - h_1) + \tfrac{1}{2}(c_2^2 - c_1^2) + g(z_2 - z_1)]\left[\frac{\text{J}}{\text{s}}\right]$$

This can immediately be reduced to:

$$\dot{Q} - \dot{W} = \dot{m}[0 - \text{calorific value fuel}]$$

All the energy flowing into the system due to the mass flow of the fuel is calculated from the calorific value:

$$\dot{Q} - \dot{W} = 6.12\left[\frac{\text{kg}}{\text{h}}\right] \times \frac{1}{3600}\left[\frac{\text{h}}{\text{s}}\right] \times 42 \times 10^6 \left[\frac{\text{J}}{\text{kg}}\right]$$

$$= -71\,400\,\text{J/s}$$

We shall use the suffix 1 with the $\dot{W}_1$ to represent initial work rate before the hill. The detail of what $\dot{Q} - \dot{W}_1$ are is not important, although the majority of this is the work done per second by the car overcoming the resistance to movement.

Consider the car climbing the hill now. We will assume that the $\dot{Q} - \dot{W}_1$ term remains constant.

$$(\dot{Q} - \dot{W}_1) - mg\dot{h} = \left(71\,400\left[\frac{\text{J}}{\text{s}}\right] + \text{increase in fuel used}\right)$$

The $mg\dot{h}$ is the extra work rate due to the hill climb; it is the weight, mg, multiplied by the distance h per unit time, $\dot{h}$.

$$mg\dot{h} = 1400\,[\text{kg}] \times 9.81\left[\frac{\text{m}}{\text{s}^2}\right] \times 0.5\left[\frac{\text{m}}{\text{s}}\right]$$

$$= 6867\,\text{J/s}$$

The increase in the fuel used is due to the extra work rate, 6867 J/s. This can now be calculated as a percentage of the original used:

$$\text{increase in fuel} = \frac{6867\,[\text{J/s}]}{71\,400\,[\text{J/s}]} \times 100\% = \mathbf{9.6\%}$$

If you look at the SFEE and think about a process with no mass flow you will see how the non-flow energy equation is derived: $Q = W + \Delta U$. The principle of conservation of energy still applies. No energy is passed in or out of the system due to mass flow, just due to heat or work crossing the boundary, the difference between the two resulting in a change in the internal energy.

6.5 Liquids

Hydraulics is the study of liquids in engineering. The word 'hydraulics' is commonly used to refer to high pressure hydraulic power systems, such as braking systems or hydraulic power assisted steering, but the term is really more general than that. A liquid is a fluid that will flow under gravity to take up the shape of its container. As with most things in engineering, hydraulics is divided into stationary liquids and dynamic liquids. **Hydrostatics** refers to the study of liquids at rest such as in a storage tank or some hydraulic machinery. **Hydrodynamics** refers to the study of liquids in motion, such as water flowing through a heating system. In a vehicle there are several liquid systems: for instance, the fuel system, the lubrication system, the cooling system and hydraulic power systems. All the liquids in these systems serve different purposes and the automobile engineer must understand the relationships between the properties and characteristics such as pressure, temperature, flow rate and flow profile.

Volumetric expansion of liquids

When the temperature of most substances changes, then so does the volume. You will be aware that checking liquid levels in a vehicle should be usually done when the vehicle is warmed up and not when it is cold, i.e. at the normal running temperature. This change in volume needs to be predicted when designing and performing calculations on a fluid system (Figure 6.21).

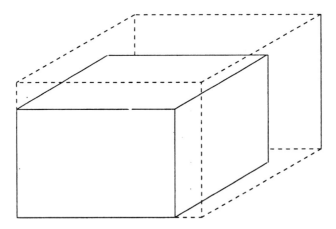

Figure 6.21 Volumetric expansion

Not all substances expand equally when subjected to the same temperature change. These changes in volume are calculated using constants or **coefficients** for a particular substance. For solids, a coefficient of linear expansion or contraction is used. It can predict the change in length of a linear dimension. Liquids take up the shape of their container and do not have fixed dimensions. The change in volume of a liquid that undergoes a temperature change is proportional to the original volume and the change in temperature and is calculated using a **coefficient of volumetric expansion**. This is the change in volume per unit volume for one degree change in the temperature of a substance. The total change in volume is calculated using the following formula:

change in volume = coefficient of volumetric expansion × original volume
× temperature rise

We can give the coefficient of volumetric expansion a symbol, γ. The Greek letter δ, pronounced delta, is usually used to represent the words 'change in'. If V is the original volume, δV is the change in volume and δT is the change in temperature, then:

change in volume, $\delta V = \gamma \times V \times \delta T$

The units of γ are expressed per degree kelvin (/K). Units check: $[m^3] = [1/K] \times [m^3] \times [K]$.

❏ *Example 6.19*

An oil tank is 2 m long by 1.5 m wide. The oil is 3 m deep in the tank. Calculate the increase in depth of the oil when the temperature of the oil increases by 10°C. Neglect any expansion of the tank material. The coefficient of volumetric expansion for the oil is 70×10^{-5} /K.

Remember that a change in temperature in degrees Celsius is equal to the change in kelvin.

Increase in volume, $\delta V = \gamma V \delta T$

$$= 70 \times 10^{-5} \left[\frac{1}{K}\right] \times 2 \, [m] \times 1.5 \, [m] \times 3 \, [m] \times 10 \, [K]$$

$$= 63 \times 10^{-3} \, [m^3]$$

$$\text{Increase in depth} = \frac{\text{increase in volume}}{\text{area of base}}$$

$$= \frac{63 \times 10^{-3} \, [m^3]}{2 \, [m] \times 1.5 \, [m]}$$

$$= 21 \times 10^{-3} \, m = \textbf{21 mm}$$

Mixing liquids of different densities

When mixing liquids of different densities together, you can assume that the volumes and masses are not affected due to the mixing. So if two liquids are mixed together, the final mass of the mixture is equal to the two masses added together. In a similar manner, the final volume of the mixture is equal to the two volumes added together. The density of the mixture then equals the total mass divided by the total volume:

$$\text{Density of the mixture} = \frac{\text{total mass}}{\text{total volume}}$$

☐ *Example 6.20*

2 kg of oil of density $\rho = 850 \, \text{kg/m}^3$ are mixed with 5 kg of oil of density $\rho = 910 \, \text{kg/m}^3$. Find the density of the final oil.

As $\rho = \dfrac{m}{V}$ then, $V = \dfrac{m}{\rho}$

$$\text{Total volume, } V = \left(\frac{2}{850} + \frac{5}{910}\right)\left[\frac{[\text{kg}]}{[\text{kg/m}^3]}\right]$$

$$= 7.847 \times 10^{-3} \, \text{m}^3$$

$$\text{Total mass} = 5 \, [\text{kg}] + 2 \, [\text{kg}] = 7 \, \text{kg}$$

$$\text{Density of mixture} = \frac{\text{total mass}}{\text{total volume}}$$

$$= \frac{7 \, [\text{kg}]}{7.847 \times 10^{-3} \, [\text{m}^3]}$$

$$= \mathbf{892 \, kg/m^3}$$

Pascal's laws

Pascal was a scientist in the seventeenth century. Amongst other things he was responsible for developing laws on fluid pressure. For a fluid that is at rest Pascal's laws are:

1. The pressure is the same throughout the fluid if the weight of the fluid is ignored, or in other words, the pressure in a fluid is equal at the same horizontal level.
2. The pressure acts equally in all directions at the same time.
3. The pressure acts at right angles to any surface in contact with the fluid.

Notice that the word 'fluid' is used rather than 'liquid'. This is because these laws apply to gases as well as liquids.

A liquid in a closed cylinder can be put under pressure by applying a force to the piston (Figure 6.22). The pressure p in the liquid is calculated from:

$$p = \frac{F}{a}\left[\frac{\text{N}}{\text{m}^2}\right]$$

Now look at the hydraulic system in Figure 6.23. The load D at the load ram is balanced by a smaller force F at the effort ram. The pressure due to the load D is given by:

$$\text{pressure} = \frac{\text{load}}{\text{area}} = \frac{D}{A}$$

The pressure due to the force F:

$$\text{pressure} = \frac{\text{load}}{\text{area}} = \frac{F}{a}$$

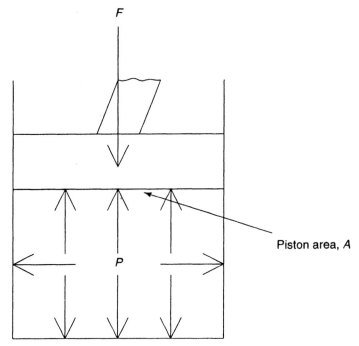

Figure 6.22

If the system is in equilibrium then the pressure is equal in both rams:

$$\frac{D}{A} = \frac{F}{a}$$

This can be rearranged to:

$$\frac{D}{F} = \frac{A}{a}$$

So, for equilibrium, the ratios of the applied forces and ram areas are equal.

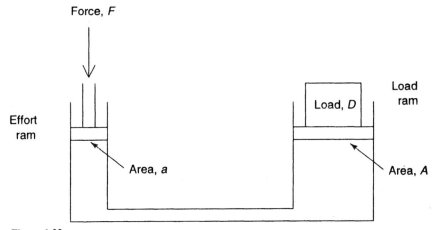

Figure 6.23

Work and power

Suppose we want to calculate the work done in raising the load D by a certain height L. Liquid must be displaced from the effort ram to the load ram. The volume V of liquid displaced will be equal to the load ram area A by the height L:

volume displaced $V = A \times L$

The work done equals the force multiplied by the distance moved:

work done $W = D \times L$

As pressure $p = \dfrac{D}{A}$

then $D = p \times A$

We can substitute this back into the equation for work:

work $W = p \times A \times L$

This is the same as pressure multiplied by volume:

$$W = p \times V$$

Units check: $[\text{J}] = \left[\dfrac{\text{N}}{\text{m}^2}\right] \times [\text{m}^3] = [\text{Nm}]$

Power is the rate of doing work, i.e. $p \times V$ per unit of time. The volume V per unit of time is the flow rate. We can represent the flow rate with $\dot{V}$, the units being $[\text{m}^3/\text{s}]$. This is the volumetric flow rate. Do not confuse this with mass flow rate $\dot{m}$ measured in kg/s.

$\therefore$ Power $= p \times \dot{V} \left[\dfrac{\text{N}}{\text{m}^2}\dfrac{\text{m}^3}{\text{s}}\right] = \left[\dfrac{\text{Nm}}{\text{s}}\right] = [\text{W}]$

❏ Example 6.21

A pump delivers 600 l/h of water against a pressure of 10 bar. Calculate the power that the pump produces.

Power $= p \times \dot{V}$

$= 10\,[\text{bar}] \times 10^5 \left[\dfrac{\text{N}}{\text{m}^2\,\text{bar}}\right] \times 600 \left[\dfrac{\text{litres}}{\text{h}}\right] \times \dfrac{1}{1000}\left[\dfrac{\text{m}^3}{\text{litres}}\right] \times \dfrac{1}{3600}\left[\dfrac{\text{h}}{\text{s}}\right]$

$= 166.667\ \text{J/s} = \mathbf{166.7\ W}$

Pressure head

Pressure in a liquid varies with depth due to the liquid's own weight. To determine the pressure in a liquid at a depth h below the surface, imagine that there is a column of liquid as shown in Figure 6.24.

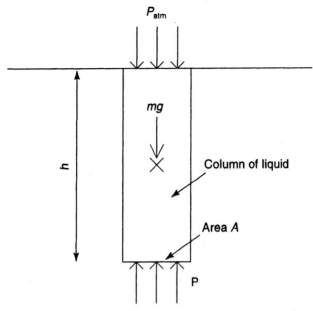

Figure 6.24

The liquid has a cross-sectional area A and its depth is h. As the liquid is at rest then all forces are in equilibrium. Horizontal forces are in complete equilibrium, as the pressure at each level is equal. Therefore we only need to consider vertical forces to calculate the pressure at the depth h.

The downward forces consist of the force due to atmospheric pressure, p_{atm}, and the force of gravity on the column of liquid:

$$\text{downward forces} = (p_{atm} \times A) + (m \times g)$$

If the density of the liquid can be found by dividing the mass by the volume, $\rho = m/V$, then the mass is equal to density multiplied by the volume, $m = \rho V$. Substituting this back into the equation we get:

$$\text{downward forces} = (p_{atm} \times A) + (\rho \times V \times g)$$

Volume is equal to cross-sectional area multiplied by height, so:

$$\text{downward forces} = (p_{atm} \times A) + (\rho \times A \times h \times g)$$

The only upward forces are due to the pressure of the liquid at the depth, h. If pressure, $p = F/A$, then the force, $F = p \times A$.

If upward forces = downward forces, then

$$(p \times A) = (p_{atm} \times A) + (\rho \times A \times h \times g)$$

The area A is a multiple of both sides of the equation so it can be cancelled out, leaving:

$$p = p_{atm} + (\rho \times h \times g)$$

The pressure p on the left side of the equation is the absolute pressure at the depth h

in the liquid. If we consider this to be the pressure above atmospheric pressure (i.e. gauge pressure), rather than the absolute pressure, then the atmospheric pressure term p on the right side of the equation can be ignored. Remember that absolute pressure is equal to atmospheric pressure plus the pressure above atmospheric pressure, i.e. the gauge pressure, $p_{abs} = p_{atm} + p_g$. This leaves:

$$p = \rho g h$$

Units check: $\left[\dfrac{N}{m^2}\right] = \left[\dfrac{kg}{m^3}\right]\left[\dfrac{m}{s^2}\right][m] = \left[\dfrac{kg}{m\,s^2}\right]$

We now need to convert the kg into other units.

As $1\,N = 1\,kg\,m/s^2$ so $1\,kg = Ns^2/m$

Substituting this back into the equation:

$$\left[\dfrac{kg}{m\,s^2}\right] = \left[\dfrac{Ns^2}{m^2\,s^2}\right] = \left[\dfrac{N}{m^2}\right]$$

It is worth working this out yourself.

We can assume that, for any liquid, the density ρ and acceleration due to gravity g will remain constant. Therefore, if the pressure at depth $h = \rho g h$, then the pressure is proportional to the depth h (i.e. $p \propto h$). The depth h is sometimes called **head** in engineering. Pressure is sometimes expressed in terms of head. For instance, the pressure a pump has to produce may be quoted in metres of water. As a rough guide 1 bar gauge ($= 100\,000\,N/m^2$) is equal to approximately 10 m head of water.

☐ Example 6.22

Calculate the pressure at the base of a vertical pipe 42 m high and of diameter 0.1 m, when it is filled with oil of density 880 kg/m^3.

$$p = \rho g h$$

$$= 880\left[\dfrac{kg}{m^3}\right] \times 9.81\left[\dfrac{m}{s^2}\right] \times 42\,[m]$$

$$= 362.578\ kN/m^2$$

Notice that the pipe diameter has nothing to do with the pressure.

☐ Example 6.23

Calculate the maximum theoretical height that water can be sucked up a pipe by a pump when the atmospheric pressure is 1.01 bar.

When a pump sucks fluid up a height then this is referred to as a **suction lift**. The pump reduces the pressure at the inlet port below atmospheric pressure and atmospheric pressure acting on the fluid forces the fluid up towards the pump. If the pump could create a perfect vacuum, then the maximum pressure available to push it up the inlet pipe is atmospheric pressure. If a perfect vacuum is created at the pump inlet, then for equilibrium at the base of the suction pipe of cross-sectional area A:

upward forces = downward forces

As pressure, $p = \dfrac{F}{A}$ then force, $F = p \times A$:

$$p_{atm} \times A = \rho \times g \times h \times A$$

$$\therefore h = \frac{p_{atm}}{\rho g}$$

$$= \frac{1.01\,[bar] \times 10^5 \left[\dfrac{N}{m^2}\right]}{1000 \left[\dfrac{kg}{m^3}\right] \times 9.81 \left[\dfrac{m}{s^2}\right]}$$

$$= 10.296 \text{ m}$$

A manometer is a device for measuring low pressures. It consists of a U-tube containing either water or mercury, depending upon the pressure range to be measured. Mercury will measure higher pressures than water because of its higher density. Figure 6.25 shows a manometer. The two ends of the tubes can be connected to two different pressures. The difference in the height in each leg indicates the pressure difference. If one end of the tube is open to the atmosphere and the other is connected to the pressure to be measured, then the difference in height will indicate the pressure above atmospheric, i.e. gauge pressure. The usual head formula is used: $p = \rho g h$.

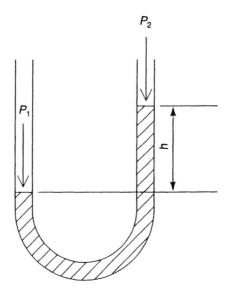

Figure 6.25 Manometer

❏ *Example 6.24*

The pressure in a gas cylinder is measured by attaching a suitable water U-tube manometer. The difference in liquid level is 0.721 m. Calculate the gauge pressure.

$$p = \rho gh$$

$$= 1000 \left[\frac{kg}{m^3}\right] \times 9.81 \left[\frac{m}{s^2}\right] \times 0.721 \, [m]$$

$$= 7073.01 \, \text{N/m}^2 = \mathbf{7.073 \, kN/m^2}$$

Liquid flow

When a solid body moves, however complex the motion is, the individual particles of the body all move in the same direction and in a similar manner. With a liquid, all the individual particles are not fixed relative to each other and the individual particles can all move separately. This can make the motion of a volume of liquid difficult to describe.

Laminar and turbulent flow

When a liquid flows along a pipe, the motion is resisted by **viscosity**. Viscosity is the resistance to the rate of change of shape. An example of a liquid with a high viscosity is bearing grease. An example of a liquid with a low viscosity is petrol. The viscosity causes a pressure drop along the length of the pipe. This requires energy to drive the liquid through the pipe. The resistance to flow also depends upon the type of flow. As the liquid flows past the pipe walls, the particles nearest to the walls are slowed down due to frictional forces between the solid and the liquid. These frictional forces are called viscous forces. The effect of the liquid slowed down like this is called **viscous drag**. When the velocity of the flow is low, the individual particles of the liquid all move in the same direction parallel to the sides of the pipe walls. The flow profile can be shown by a cross-section of the liquid and the pipe, using arrow lengths to represent the liquid velocity.

This type of flow is shown in Figure 6.26 and is called **laminar flow**, or **streamline flow**. Laminar flow in a pipe is defined as liquid flow where all the particles at a given radius from the pipe centre all move at the same velocity, the velocity being highest at the centre and lowest at the pipe walls. At higher flow velocities, the motion becomes chaotic. The individual particles no longer flow in the same direction, but move

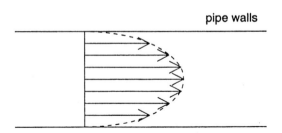

Figure 6.26 Laminar flow

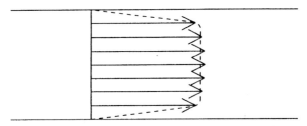

Figure 6.27 Turbulent flow

around in all directions at random in the general direction of the liquid motion. This type of flow is called **turbulent flow**. The flow pattern is less ordered (Figure 6.27).

The viscosity of the liquid affects the type of flow as well as the velocity. At low velocities and high viscosities the flow tends to be laminar. At high velocities and low viscosities the flow tends to be turbulent. For example, you would expect slow oil flow to be laminar and fast water flow to be turbulent. Other things that affect the type of flow are the pipe diameter and liquid density.

To be able to calculate the volume flow rate through a pipe we have to make some assumptions. If the flow is assumed to be **perfect** the problem is simplified. Perfect flow means that every particle of the liquid flows with the same velocity in straight lines parallel to the walls of the pipe. This type of flow does not exist in reality but is another useful ideal for simplifying calculations (Figure 6.28).

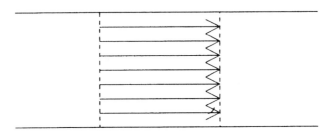

Figure 6.28 Perfect flow

Volumetric flow rate and mass flow rate

The velocity of flow of a unit volume of fluid through a pipe can now be considered. This can be found by recording the quantity of liquid flowing through a system in a given time. The average flow rate can be calculated as:

$$\text{average flow rate, } \dot{V} = \frac{\text{volume of liquid}}{\text{time}} \left[\frac{m^3}{s}\right]$$

$$= \frac{V}{t}$$

If the average flow rate is then divided by the cross-sectional area of the pipe, the average velocity is found:

$$\text{average velocity } c = \frac{\text{average flow rate}}{\text{cross-sectional area}} = \frac{p}{A}$$

We use c to represent velocity rather than v so there is no confusion with volume V or specific volume v.

Units check: $\dfrac{[m^3/s]}{[m^2]} = \left[\dfrac{m}{s}\right]$

The mass flow rate can be found from the volumetric flow rate. Remember that density is equal to mass divided by volume, $\rho = m/V$ and so $m = \rho \times V$.

If volumetric flow rate $\dot{V} = c \times A$

Mass flow rate $\dot{m} =$ volumetric flow rate $\times$ density
$$= \dot{V} \times \rho$$
$$= c \times A \times \rho$$

Equation of continuity

Consider liquid flowing through a pipe that reduces in diameter over its length between points 1 and 2 (see Figure 6.29).

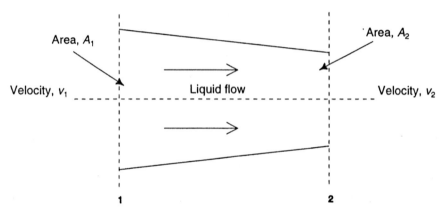

Figure 6.29 Flow through a tapering pipe

The mass flow rate passing any section is constant. The cross-sectional areas at sections 1 and 2 are A_1 and A_2, and the fluid velocities are c_1 and c_2, respectively. As the mass flow rate is constant, we can say that:

$$c_1 \times A_1 \times \rho = c_2 \times A_2 \times \rho$$

The density will remain constant and can be cancelled out:

$$c_1 \times A_1 = c_2 \times A_2$$

We can rearrange this to give:

$$\frac{c_1}{c_2} = \frac{A_2}{A_1}$$

When the pipe dimensions are known the ratio of the velocities can be found. This is called the **equation of continuity**, and states that the velocity of the flow is inversely proportional to the area of the pipe section.

Also, as $A_1 = \dfrac{\pi d_1^2}{4}$ and $A_2 = \dfrac{\pi d_2^2}{4}$ then:

$$\frac{c_1}{c_2} = \frac{\pi d_2^2}{4} \times \frac{4}{\pi d_1^2} = \frac{d_2^2}{d_1^2}$$

$$\therefore \quad \frac{c_1}{c_2} = \frac{A_2}{A_1} = \frac{d_2^2}{d_1^2}$$

So, not only is the velocity inversely proportional to the area of the pipe section, it is also inversely proportional to the square of the pipe diameters. For example, if the diameter of the pipe were to reduce by one-half then the velocity would increase to four times its original value.

☐ *Example 6.25*

The diameter of a length of tapered pipe reduces from 0.25 m to 0.12 m over a length of 1.5 m. Liquid flows from the larger diameter to the smaller diameter. The velocity of the liquid is 2.5 m/s at the inlet to the pipe. What is the velocity at the outlet of the pipe?

$$\frac{c_1}{c_2} = \frac{d_2^2}{d_1^2}$$

$$\therefore c_2 = c_1 \times \frac{d_1^2}{d_2^2}$$

$$= 2.5 \left[\frac{m}{s}\right] \times \frac{0.25^2 \, [m^2]}{0.12^2 \, [m^2]}$$

$$= 10.851 \text{ m/s}$$

Bernoulli's equation

Bernoulli's equation states that when an incompressible liquid flows between two points in a system, the total energy at point one is equal to the total energy at point two. The total energy is composed of:

1. potential energy
2. pressure energy or flow work
3. kinetic energy.

$$\text{Potential energy} = mgh$$
$$\text{Pressure energy or flow work} = pV$$
$$\text{kinetic energy} = \frac{1}{2} mc^2$$

These were defined in the chapter on the steady flow energy equation. Inserting these values into the equation:

$$mgh_1 + p_1V + \frac{1}{2}mc_1^2 = mgh + p_2V + \frac{1}{2}mc_2^2$$

We could have derived this from the steady flow energy equation. A section of the pipe can be considered as a system with a steady mass flow rate in and a steady mass flow rate out. The steady flow energy equation is:

$$\dot{Q} + \dot{m}\left(u_1 + gh_1 + \frac{1}{2}c_1^2 + p_1v_1\right)\left[\frac{J}{s}\right] = \dot{m}\left(u_2 + gh_2 + \frac{1}{2}c_2^2 + p_2v_2\right) + \dot{W}\left[\frac{J}{s}\right]$$

This equation considers the energy flow rate entering or leaving the system per second. If energy flow rate entering or leaving the system is considered during one second the equation becomes:

$$Q + m\left(u_1 + gh_1 + \frac{1}{2}c_1^2 + p_1v_1\right)[J] = m\left(u_2 + gh_2 + \frac{1}{2}c_2^2 + p_2v_2\right) + W[J]$$

Through a section of pipe, we could assume no heat transfer and no work transfer through the pipe wall. We could also assume that there are no internal energy changes between the liquid entering the system and leaving the system. The equation becomes:

$$m\left(gh_1 + \frac{1}{2}c_1^2 + p_1v_1\right)[J] = m\left(gh_2 + \frac{1}{2}c_2^2 + p_2v_2\right)[J]$$

which is the same as Bernoulli's equation:

$$mgh_1 + p_1V + \frac{1}{2}mc_1^2 = mgh_2 + p_2V + \frac{1}{2}mc_2^2$$

In the SFEE equation remember z is used in place of h for height to avoid confusion with enthalpy.

As this is applied to hydraulic systems, it is usually expressed in terms of pressure head h, i.e. divide throughout by mg:

$$h_1 + \frac{p_1V}{mg} + \frac{c_1^2}{2g} = h_2 + \frac{p_2V}{mg} + \frac{c_2^2}{2g}$$

As density, $\rho = m/V$ then $V/m = 1/\rho$. This can be substituted into the equation to give an alternative form:

$$h_1 + \frac{p_1}{\rho g} + \frac{c_1^2}{2g} = h_2 + \frac{p_2}{\rho g} + \frac{c_2^2}{2g}$$

This assumes that there is a perfect interchange of energy. No account is taken of the 'losses' of energy between points 1 and 2 in the system, e.g. heat transfer through the pipe walls. These losses can be allowed for in Bernoulli's equation with a little modification.

Total energy at 1 = total energy at 2 + total energy 'losses'. between 1 and 2

$$h_1 + \frac{p_1}{\rho g} + \frac{c_1^2}{2g} = h_2 + \frac{p_2}{\rho g} + \frac{c_2^2}{2g} + \text{losses}$$

☐ *Example 6.26*

Diesel fuel of density 900 kg/m^3 flows down a pipe. The difference in height over the length is 1.5 m. The pressure and velocity of the oil at the top of the pipe are 170 kN/m^2 and 5 m/s, respectively. The velocity of the oil at the bottom of the pipe is 15 m/s. Calculate the pressure at the bottom of the pipe. Assume that there are no losses.

We'll call the top of the pipe point 1 and the bottom of the pipe point 2.

$$h_1 + \frac{p_1}{\rho g} + \frac{c_1^2}{2g} = h_2 + \frac{p_2}{\rho g} + \frac{c_2^2}{2g}$$

Rearranging for p_2:

$$p_2 = \left(h_1 + \frac{p_1}{\rho g} + \frac{c_1^2}{2g} - h_2 - \frac{c_2^2}{2g} \right) \rho g$$

$$= \left(1.5 + \frac{170 \times 10^3}{900 \times 9.81} + \frac{5^2}{2 \times 9.81} - 0 - \frac{15^2}{2 \times 9.81} \right) [\text{m}] \times 900 \left[\frac{\text{kg}}{\text{m}^3}\right] \times 9.81 \left[\frac{\text{m}}{\text{s}^2}\right]$$

$$= (1.5 + 19.255 + 1.274 - 0 - 11.478) [\text{m}] \times 900 \left[\frac{\text{kg}}{\text{m}^3}\right] \times 9.81 \left[\frac{\text{m}}{\text{s}^2}\right]$$

$$= \mathbf{93.155 \text{ kN/m}^2}$$

Problems 6

1. The temperature of a room is 21°C. What is this in absolute units?
2. A copper vessel of mass 0.5 kg contains 2.5 kg of water. The initial temperature of the water and copper is 16°C. The vessel is heated until the final steady temperature is 40°C. Ten per cent of the energy supplied is lost to the atmosphere. How much energy is required? The specific heat capacity for the copper $c = 0.391 \text{ kJ/kg}$. For the water, the specific heat capacity $c = 4.18 \text{ kJ/kg}$.
3. The pressure of a tank is measured with a mercury manometer. The reading is 900 mm of mercury. Atmospheric pressure at the time of the reading is 1.005 bar. What is the absolute pressure in the tank? The density of mercury is $13\,500 \text{ kg/m}^3$.
4. The pressure of air in a pressure bottle is measured with a water manometer. The reading is 850 mm. The atmospheric pressure at this time is 759 mm of mercury. Calculate the absolute pressure of the air. The density of water is 1000 kg/m^3. The density of mercury is $13\,500 \text{ kg/m}^3$.
5. 5 kg of steam that is a dry saturated vapour is converted to water at 20°C. How much energy is obtained? The specific heat capacity is $c = 4.18 \text{ kJ/kg}$ and the specific latent heat of vaporisation is 2260 kJ/kg.

6. The cooling system of a vehicle contains 9 kg of water at a temperature of 45°C. How much energy is required to evaporate 20% of the water at atmospheric pressure? The specific heat capacity of the water is $c = 4.18$ kJ/kg and the specific latent heat of vaporisation of the water is 2261.8 kJ/kg.

7. The internal energy of a system decreases by 1500 kJ when 550 kJ of heat is transferred to the system from the surroundings. What is the work transfer?

8. One kilogram of a fluid at a pressure 200 kN/m^2 is compressed from 0.6 m^3 to 0.15 m^3. What is the work done?

9. A gas with a volume of 0.05 m^3 and an absolute pressure of 500 kN/m^2 expands until the volume is 0.15 m^3. The temperature remains constant throughout the process. Calculate the final pressure.

10. The compression ratio of an engine is 10 to 1. The pressure at the beginning of the compression is 50 kN/m^2. The pressure at the end of the compression is 3000 kN/m^2. At the start of the process the air is 30°C. What is the final temperature?

11. A wall is double-skinned. The outer layer is brick 150 mm thick with a thermal conductivity of 0.85 W/m K. The inner layer is made of an insulation material, 75 mm thick, and with a thermal conductivity of 0.07 W/m K. The temperature inside is 43°C and the temperature outside is 2°C. Calculate the heat loss through the wall.

12. Describe the meaning of the terms 'emissivity' and 'black body'. What type of surfaces absorb the most radiation?

13. Air in a cylinder has a pressure of 50 kN/m^2, a temperature of 30°C and a volume of 0.09 m^3. What is the mass of air? The specific gas constant R for the air is 0.287 kJ/kg K.

14. Cooling water flows through a system at a rate of 13 kg/min. Calculate the velocity when it flows through a pipe of diameter 30 mm and a pipe of diameter 46 mm. The density of the cooling water is 1015 kg/m^3.

15. A pump discharges water of density 1000 kg/m^3 through a 50 mm diameter pipe at 20 litres/min, to a height of 1.5 m. The efficiency of the pump is 70%. Find the power required by the pump.

16. A gas enters a system at a temperature of 65°C, a pressure of 125 kN/m^2 and a velocity of 7.7 m/s through a pipe of diameter 0.1 m. The gas leaves the system at a temperature of 175°C, a pressure of 225 kN/m^2 through a pipe diameter 0.3 m. Calculate the velocity of the gas leaving the system. The specific gas constant R for the gas is 0.292 kJ/kg K.

17. Thirty-five cubic metres per hour of fresh water flows vertically upwards through a tapering pipe and discharges to the atmosphere. The flow area of the pipe exit is 5500 mm^2. The flow area of the pipe base is 3200 mm^2. The height of the top of the pipe is 2 m above the base. Apply Bernoulli's equation in the equivalent head form to find the gauge pressure at the base of the pipe. Assume the density of the water to be 1000 kg/m^3.

18. A car radiator is completely filled with cold water; this requires 12 litres. The car travels a short journey. During the journey 0.25 litres of water is lost through the overflow pipe. The initial temperature of the water is 10°C. Find the final temperature of the water. The coefficient of volumetric expansion of the water is 43×10^{-5} /K.

19. Find the density of an oil if a vertical column of this oil 1.5 m high can be balanced by a vertical column of mercury 0.105 m high. Assume the density of the mercury to be 13 600 kg/m^3.

20. Fresh water flows through a short horizontal tapering pipe, from the larger diameter to the smaller diameter. At the large end the velocity is 2 m/s. The pressure reduces by 15 kN/m^2 through the pipe. Calculate the velocity at the smaller end.

7 | Control and instrumentation

7.1 Modelling and control systems

What is a system?

So far, we have looked at mechanical and thermodynamic systems. We have seen that a system is a collection of connected components or matter within a boundary that we want to investigate. The boundary is an imaginary line around the components to make the study of them easier. The components will all interact with each other in some way and the boundary shows the inputs and outputs of a system as any signals that cross it. A system can be just about anything, e.g. an arm, an engine or a power station. It is the region in space that is of interest. Everything outside of the boundary is called the environment. The signals that cross the boundary from the environment are called **system inputs** and the signals that cross the boundary from the system are called **system outputs**. It is the position of the boundary that defines the system.

Consider a cooling water system that uses an electric motor driven pump to provide the water pressure (Figures 7.1 and 7.2). The speed of the pump N is

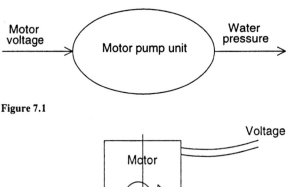

Figure 7.1

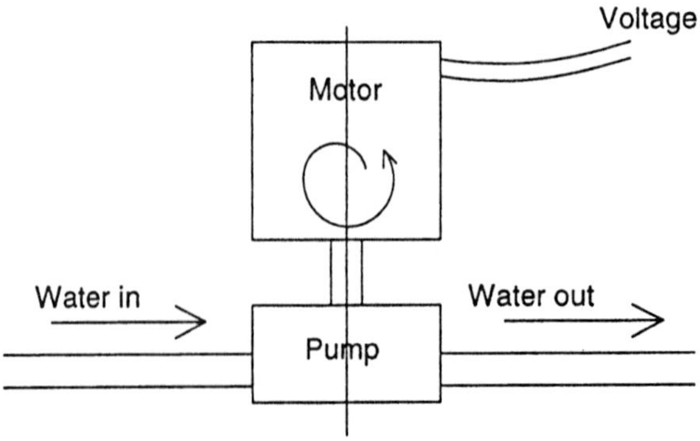

Figure 7.2

proportional to the water pressure p, i.e. $p \propto N$. The voltage supply V to the motor is proportional to the speed of the motor and pump N, i.e. $V \propto N$. Therefore the voltage applied is proportional to the water pressure, $V \propto p$. The voltage is the input to the system. The output is the water pressure. We could however have positioned the boundary differently.

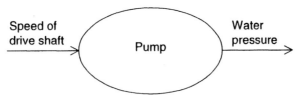

Figure 7.3

In Figure 7.3 the pump is the system. The input is the speed of the drive shaft N and the output is the water pressure p, again. Often, there will only be one input that varies and only one output that is affected by this varying input. As in the above example, these are the easiest to analyse and the ones we will look at here. A system may not necessarily be a physical system, like that above: it could concern economics or management for instance. We will restrict our studies to physical systems.

Why model a system?

A model of a system, like any model, is a representation of that system. The model is used to predict how the system will respond to certain inputs. When a driver gets into an unfamiliar vehicle, the controls may feel strange and driving may seem difficult at first. Quickly, however, the driver gets used to the vehicle, to the response of the engine revolutions from the accelerator pedal, to the response of the car when the steering wheel is turned, to the deceleration on braking, etc. This happens because the driver learns about the vehicle, i.e. a model is developed in the driver's mind. This model is a 'picture' of the performance of the vehicle and allows the response of the vehicle to be predicted.

The models that engineers often use are mathematical models. These can be developed from equations and are used to analyse system performance or control. It is useful to be able to predict how a system will perform under different conditions. The relationship of interest is between the variable input and measured output, such as in the pump example above.

The model is only an approximation of the system though, and certain assumptions have to be made to simplify the performance.

Block diagrams

Having decided that you want to model a system, what now? We have to look at what exactly makes up that system and how it interacts with the environment. An imaginary boundary has been drawn around the system. We now need to represent the system in a simple way. A common method of doing this is using a *block diagram*. The individual functioning parts of a system can be represented with blocks, and arrows between the blocks drawn to show how the parts are related. The arrows show the signal flow through the system. Each block represents a relationship between its

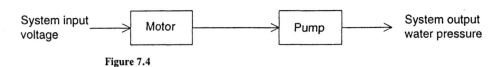

Figure 7.4

input and output. The motor pump example above can be represented as shown in Figure 7.4.

The usual convention is to draw the system input at the left side of the page and the system output at the right side. The lines between the block are signals and the arrows show the direction of the signal flow. For instance, the arrow between the motor and the pump is the shaft speed, i.e. the output from the motor is the shaft speed and the input to the pump is the shaft speed. This signal is a precise variable. It is not simply the motion of the shaft: it is the speed of the shaft that is proportional to the water pressure and so it is the pump block that represents the relationship between shaft speed and water pressure.

Subsystems

A system often consists of several interacting subsystems. How complex the model needs to be will depend upon what information you require from it. Consider a petrol engine as shown in Figure 7.5.

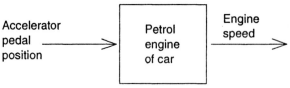

Figure 7.5

The input to the engine is the accelerator pedal position. The output is the engine speed. The engine can be considered to consist of four subsystems: the carburettor, the intake manifold, the combustion and the engine dynamics (Figure 7.6).

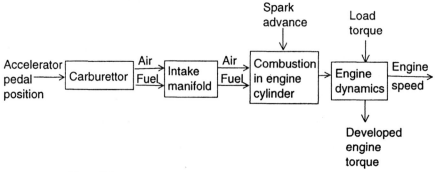

Figure 7.6

The spark advance is also a controllable input. The load torque is a disturbance input. The developed torque is also indicated as an output. Each subsystem can be investigated further, if necessary. A model outline of the carburettor could look like Figure 7.7.

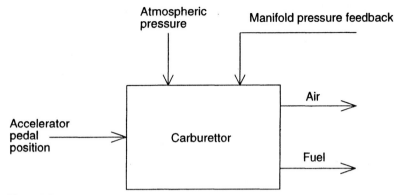

Figure 7.7

The accelerator pedal position is the controllable input. The outputs are the air and fuel, which become the inputs to the intake manifold. The atmospheric pressure and the manifold pressure affect the operation of the carburettor and are shown as disturbance inputs. There is a feedback signal of the manifold pressure from the intake manifold subsystem.

Mathematical models

The mathematical model of a system describes the relationship between the inputs and the outputs. Mathematical laws and formulae can be used to develop a model. The variables, the theories and laws, and the assumptions made, all affect the model developed.

❏ *Example 7.1*

Consider a simple coil spring that has one end fixed against a surface as shown in Figure 7.8.

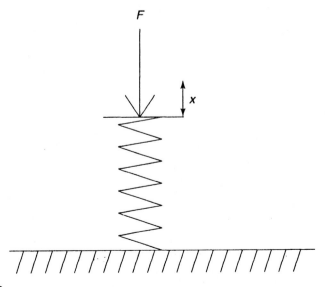

Figure 7.8

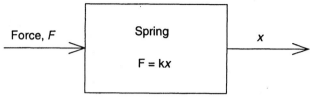

Figure 7.9

A force, F, is applied to the other end. The mass of the spring can be assumed to be negligible. The spring will compress by distance x. A model is required that can predict the deflection x for an input force F. We could represent this by a block that has an input of F and an output of x, as shown in Figure 7.9.

If the spring stiffness is k then the force applied from the deflection is: $F = k \times x$. This is the equation relating the output to the input of the block and we now have a simple mathematical model of the system. This can be used to predict the deflection x when a force is applied.

The decision that we made, to the mass of the spring, simplifies the model. Had we not ignored the mass of the spring then the mathematical model would have been more complex and more accurate, incorporating the effects of gravity. We must also assume that the force is not so large that the coils of the spring are pushed together and the above formula would no longer apply. How complex the model is made depends on its purpose.

One thing to notice is that the model is not time-dependent. This means that, for a constant input (force), the output remains constant with time. This is not always the case.

☐ *Example 7.2*

Consider a hydraulic damper. A piston can slide in a cylinder and the cylinder is filled with hydraulic oil. There are holes in the piston so that the oil can flow through it as it slides. A model is required that predicts the position of the piston when a force is applied.

We will assume that the piston has a negligible mass and that the oil can flow freely through the piston so that the pressure remains equal each side of the piston. When the force is applied the piston will move. The oil in the cylinder creates a drag on the piston as it moves.

Look at Figures 7.10 and 7.11. The velocity of the piston is proportional to the viscous drag on the piston (see page 240). We know the relationship between the force and the velocity but not between the force and the displacement x. The relationship between the force applied and the velocity v of the piston is

$$F = Cv$$

$$= C \frac{dx}{dt}$$

where C is a constant.

The velocity is represented as dx/dt meaning the change in the displacement x occurring in a certain time t (see page 17). Velocity is represented like this rather than

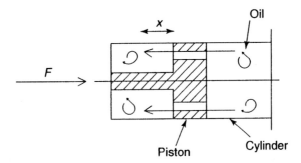

Figure 7.10

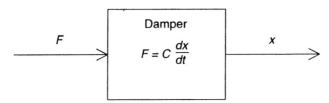

Figure 7.11 First order response

by a single letter such as v to highlight the relationship between the force F and the position x. The difference between this example and the last one with the spring is that there is not a single constant output (displacement x) for a fixed input (force). Here, the output changes with time: it is time-dependent. A time-dependent system is called a **dynamic system**. The output of a dynamic system depends on the input and time. The example here is known as a first order system, as it involves the first derivative of the output x, i.e. dx/dt. A graph of the output x measured over a period of time for a step input of F would be as shown in Figure 7.12.

This is a typical response of a first order system to a step input. Notice that there is no overshoot before the output settles at the final value.

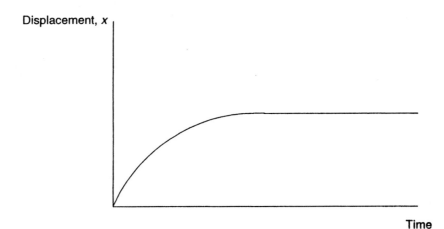

Figure 7.12

❏ *Example 7.3*

If a force is applied to a block with a mass m that stands on a frictionless surface, the block will accelerate according to Newton's second law ($F = ma$). The block is guided so that it can only move in the direction of the force. We require a model that predicts the position of the block when a force is applied.

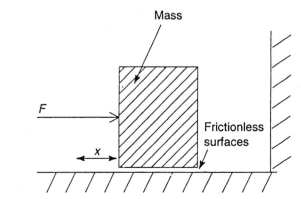

Figure 7.13

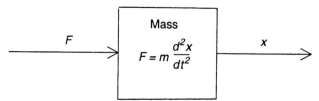

Figure 7.14

See Figures 7.13 and 7.14. We know the relationship between the force and the acceleration but not the relationship between the force and the position, or even the velocity. In Chapter 4 it was shown that velocity could be calculated and represented by dx/dt, meaning the change in displacement occurring in a time t. This is the rate of change of the displacement. If the velocity, dx/dt, is differentiated again, we obtain an expression for acceleration, d^2x/dt^2. This is the change in velocity occurring in a time t which is the rate of change in velocity. The expression relating force to acceleration can be written like this:

$$F = ma \Rightarrow F = m\frac{d^2x}{dt^2}$$

This form of the expression looks a little more complex but it now shows the relationship between the force and the displacement x. Again the model is time-dependent. This is known as a second order system, since the expression involves the second derivative of the output, i.e. d^2x/dt^2.

☐ *Example 7.4*

The above three examples can be combined into one system. See Figures 7.15 and 7.16.

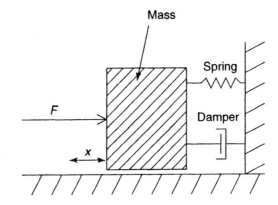

Figure 7.15

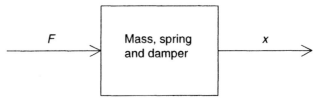

Figure 7.16

This is a common arrangement in engineering. A suspension spring and damper of a vehicle behave in a similar manner to this. To develop a model we will make the same assumptions that we made for each individual model, i.e. that all mass is negligible except for the mass of the block; the block can only move in the direction of the force and there is no friction between the block and the surface. We can apply Newton's second law of motion again:

$F = ma$

Now, however, there is more than the one input force to consider. The input force is opposed by the resistance forces of the spring F_s and the hydraulic damper F_d:

total input force $= F - F_s - F_d$

This input force is used to give the mass of block m an acceleration a or $\dfrac{d^2x}{dt^2}$.

So, $F - F_s - F_d = ma$

$$= m\frac{d^2x}{dt^2}$$

We also know that $F_s = kx$ and $F_d = C\dfrac{dx}{dt}$. Substituting these expressions into the equation we get:

$$F - Kx - C\frac{dx}{dt} = m\frac{d^2x}{dt^2}$$

Rearranging this we get,

$$F = m\frac{d^2x}{dt^2} + C\frac{dx}{dt} + Kx$$

This is obviously a more complex model than the previous three examples, but the motion of the block is more complex. The output response of this system to a step input could be as shown in Figure 7.17.

This is a typical response of a second order system to a step input. Notice that the output overshoots and oscillates before settling at its final value.

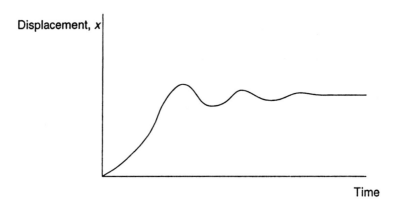

Figure 7.17

Analogous systems

Analogous systems are different systems that have a similar relationship between outputs and inputs. For example, consider the water tank shown in Figure 7.18.

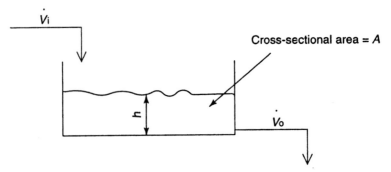

Figure 7.18

The volumetric flow of water in is $\dot{V}_i$, the volumetric flow of water out is $\dot{V}_o$ and the net flow of water in is $\dot{V}_{net}$, so that $\dot{V}_{net} = V_i - V_o$. The depth of the water in the tank is h. The relationship between the net flow in and the depth is:

$$\dot{V}_{net} = A \frac{dh}{dt}$$

where A is the cross-sectional area of the tank. This is a first order response, as it involves a first derivative. This can be compared to the hydraulic damper model with an input of force F and an output of displacement x:

$$F = C \frac{dx}{dt}$$

The difference between the two relationships is in the constants C and A. These two systems are known as **analogous systems.** Analogous systems are a useful concept. The response of one system can be simulated by another. Electrical circuits are sometimes designed as analogous systems to have comparable relationships between, for instance, input voltage and output voltage. If a controller is being designed to control, for example, the level of a large industrial boiler by varying the water pump speed, then it is far simpler and safer to test the controller initially on an analogous electrical circuit that can mimic the relationship between the input and the output of the boiler.

Control systems

A control system is used when a quantity or variable (the output) in any equipment needs to be maintained or altered in some way. This is done by adjusting one or more of the inputs. Consider a gas-powered air heater (Figure 7.19).

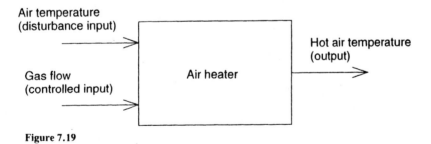

Figure 7.19

The temperature of the hot air produced by the heater can be controlled by adjusting the gas flow that is burnt. The temperature of the heated air is also affected by the temperature of the cold air entering the heater from the atmosphere. Only the gas flow can be adjusted. This is called a **controlled input.** Inputs that affect the output but cannot be adjusted are called **disturbance inputs.** The variable that is chosen as the output is not always a good indication that the system is controlled properly. The air heater measures the outlet temperature of the air, but it is probably the temperature of the air in the heated room that is of interest to the people in it.

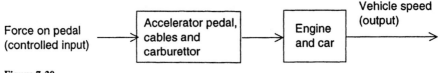

Figure 7.20

Now think about driving a car as a system (Figure 7.20). The speed of the car is controlled by adjusting the force on the accelerator pedal. There are other disturbance inputs such as the gradient of the road (not shown here). If the driver wants to maintain a speed of 35 km/h (system output), then this can be done by holding the accelerator pedal in a fixed position (system input), provided that there are no disturbances. If there is a disturbance such as a hill then speed will reduce (e.g. to 20 km/h). The driver can read the speed of the car with the speedometer. There is now a difference between the desired speed and the actual speed and the driver can apply an increased pressure on the accelerator pedal to correct the difference. This control system can be represented by the block diagram in Figure 7.21.

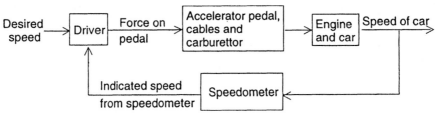

Figure 7.21

The control system takes the actual output, feeds it back to the input and compares it with the desired output. Corrective action can then be taken to reduce the difference or error. The information flows in a loop around the system. This type of control system is called a **closed loop** system. In contrast, a system such as that shown by the air heater diagram, which does not have a feedback of information, is called an **open loop** system. A system that requires a human operator to make the comparison between the system output and the desired output is a manual system. Many systems do not use a human operator, when for instance the system dynamics are fast and complex; these are called **automatic control systems**. The general form of a block diagram for an automatic control sytstem is shown in Figure 7.22.

Here, the feedback is negative. This is a common arrangement for a control system. The feedback signal Y is subtracted from the reference signal, i.e. the desired output R, to produce an error signal, E:

$$E = R - Y$$

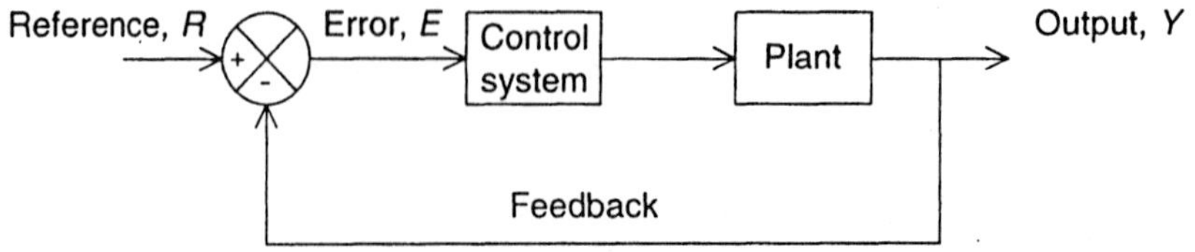

Reference, R Error, E [Control system]   [Plant]   Output, Y

Feedback

Figure 7.22

The controller is designed using a model of the system. It adjusts the input to the system in order to produce the predicted output. The actual output achieved is then compared with the desired output, i.e. the reference signal, and the error signal is produced. The controller uses the error signal to further adjust the input to the system.

Consider a heating control system for a room as shown in Figure 7.23. The output temperature of the system, i.e. the room temperature Y, is compared with the desired output temperature, i.e. the reference signal R, to produce the error signal, E. The input of the controller is the error signal E. The output of the controller is the **control effort**, applied to the heater to reduce the error signal, i.e. the difference between the system output and the desired output. The feedback is used because the actual output temperature achieved will often differ from the desired temperature. This is due to other disturbance inputs or inaccuracies of the system model to which the controller design applies.

The design of the control system is a vast branch of science and engineering all of its own and cannot be covered even briefly in a book of this size, but you should be aware of the purpose of a control system. Automatic control systems are increasingly used in motor vehicles for things such as automatic engine tuning.

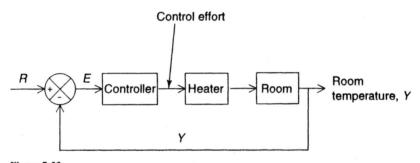

Figure 7.23

7.2 Instrumentation

Instrumentation is used for measuring quantities. These can be any quantities of a system that are of interest. In a system, we may want to measure mechanical and thermal quantities such as: strain, force, pressure, moment, torque, displacement, velocity, temperature and frequency. Electrical quantities that may be of interest are: voltage, current and resistance. We will look at mechanical and thermal measurement here. For general descriptions of instrumentation we will refer to anything that is being measured as the **process**. Quantities are measured to obtain data about the process, or in a control loop for such things as feedback. Consider all the instrumentation used in a motor vehicle. For example, the dash display may show instrument readings for: engine speed, temperature, turbocharger speed, oil pressure, vehicle speed and black ice warning. Engine speed and load are measured by the

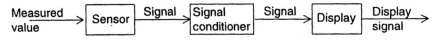

Figure 7.24 Measuring system

ignition system to alter the spark advance. Some brakes are fitted with indicators to show wear.

A measuring system consists essentially of three main components: a sensing element, a signal conditioner and a display element, as shown in Figure 7.24. The sensing element is applied to the process and in some way produces a signal that is related to the quantity to be measured. The signal converter changes the signal from the sensing element into a form suitable for the display element. The display element refers to the output from the measuring system being displayed in a form that an observer can understand. For example, a resistance thermometer uses a sensing element that, when applied to a process, produces a resistance signal that is proportional to the temperature of the process. The resistance signal can be converted into a proportional current signal. The current signal can then be used to move a pointer on a dial. Notice the signal conversion through the system:

Temperature signal → resistance signal → current signal → visual signal

Electronic instrumentation

An electronic instrumentation system is often more complex than this: the sensing element may consist of a **transducer** and a **power supply**; the signal conditioning unit may include an amplifier; the display element could consist of a data processor and a recorder. Electronic systems produce an electrical output that can be readily used for feedback in an automatic control system. If we take a more detailed look at an electronic instrumentation system, we could consider that it consists of six elements or subsystems. These are shown in Figure 7.25 in a block diagram form.

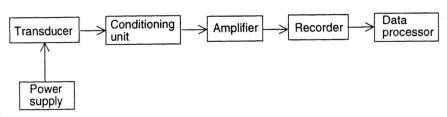

Figure 7.25 Electronic measuring system

Transducer

The transducer converts the quantity to be measured into an electrical quantity that can be monitored. The changes in the electrical quantity are proportional to the changes in the quantity measured. For example, a strain gauge is used to measure the strain of a component or structure when it is loaded. The gauge is a transducer that converts a change in strain in the component being measured into a change in electrical resistance. The change in electrical resistance is then proportional to the change in the strain.

Power supply

The transducer requires an energy source to drive it. An appropriate power supply is used to suit the transducer.

Signal conditioning

The electrical signal from the transducer may not be in a useable form. The signal conditioning unit converts or modifies the transducer output into a more useful electrical quantity. For example, the output from the strain gauge mentioned above is an electrical resistance. A voltage signal would be more useful, so the conditioning unit converts the resistance signal into a corresponding voltage signal.

Amplifiers

The voltage measuring equipment requires the signal input to be of a certain voltage. Often the output from the signal conditioning unit is a very small voltage (less than a millivolt is common). Amplifiers are used to increase the conditioning unit output voltage to a level that is suitable for reading.

Recorders

A recorder displays the voltage in a form that can be easily read. A voltmeter can be used for measuring static voltages. For dynamic voltages, i.e. voltages that vary with time, a dynamic voltmeter must be used that can record a voltage signal with respect to time. The voltage signal that represents the measured quantity is converted into a display that can be easily read visually against a suitable scale or into a digital code that can be used by the data processor.

Data processor

Data processors are used to convert the outputs from the instrument system into a form that can easily be analysed using a digital computer. They are common where large numbers of instruments are used and data is collected. Sorting out all the data would take a long time. The data processor can convert the data into graphs or tables which can be easily analysed by an engineer.

Error

Error is the difference between the measured value that instrumentation produces and the true value of the quantity measured. This must be kept to a minimum. For any instrumentation system used for measurement and analysis or for control systems, a maximum value of error must be established that will give an acceptable **accuracy** of the measured value. Errors can be due to several causes.

Accumulation of errors in each element of the instrumentation can create a large error of the overall system. A single element of an instrumentation system has a specified accuracy set by the manufacturer. If the element is used properly, in good condition and properly calibrated, then you can expect that element to operate within these accuracy limits. Accuracy is usually expressed as a percentage of full-scale deflection, e.g. a resistance thermometer with a full-scale deflection of 100°C and an accuracy quoted as ±1% means that for a reading between 0°C and 100°C, the accuracy is plus or minus 1°C. Therefore if a reading was, say, 47°C then all we can say is that the actual temperature lies between 46°C and 48°C.

The accuracy limits of the element will cause an error in the overall instrumentation system. This does not cause a problem if the error of the measured value is known and within the desired limits. However, an instrumentation system consists of several elements. Each will have its own accuracy limits that will introduce errors into the system. Even if each element operates within the manufacturer's accuracy limits, these elements can accumulate through the system. The accumulated error can then be too high to give a measured value with the necessary accuracy.

Malfunctioning of any element will cause errors in the measured value of the system. An element could malfunction because it is not properly maintained. It could also require adjusting. The **zero offset** is used to adjust the element so that when the input is zero the output is also zero. If this is not done a constant error, called a zero-offset error, exists over the whole range of the element (Figure 7.26).

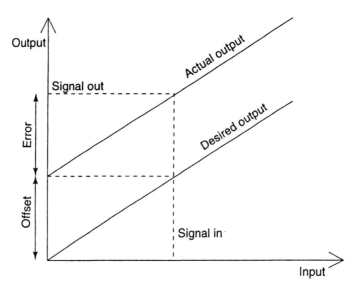

Figure 7.26 Zero offset

The element also requires **calibrating**. This adjusts the change in the output for a change in the input. When a change in the input to an element does not give the required change in the output, the error is called a calibration error. On a graph of the output against the input, this shows as an incorrect slope of the actual response (Figure 7.27).

The slope of the line is called the **sensitivity** of the instrument. If a change in the input to the element is ΔQ_i and the corresponding change of the output that this causes is ΔQ_o then:

$$\text{Sensitivity} = \frac{\Delta Q_i}{\Delta Q_o}$$

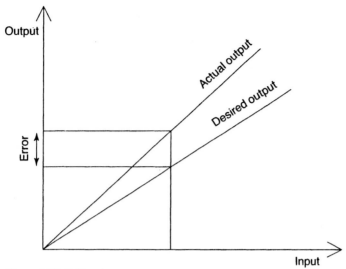

Figure 7.27 Calibration error

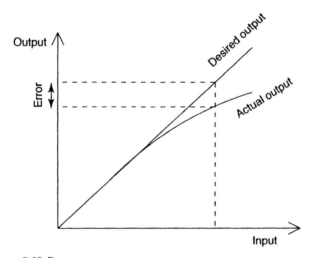

Figure 7.28 Range error

Any element will operate only over a certain range. If the input is greater than that range then the output will be subject to errors. At the upper end of the range the line on the graph of output against the input stops being straight. An input to the element above the straight part will cause **range errors** (Figure 7.28).

Transducer sensitivity to other quantities than the one being measured can result in instrumentation error. A transducer is intended to be sensitive to a particular quantity that is to be measured. It may be sensitive to other quantities. For example, a transducer may be designed to measure the oil pressure of an internal combustion engine. If the temperature changes as the measurements are being made, then errors can occur due to the transducer being sensitive to temperature as well as to the pressure of the oil. A transducer should be chosen for a system that has a negligible sensitivity to other quantities of the process that may change.

Transducer interference of the process being measured can give errors or even meaningless results. The transducer must be chosen carefully and placed on the process so that the operation of the process is not affected. The transducer may use energy from the process or apply forces to it. For example, there is no point in using a flow meter to measure the liquid flow in a pipe if, by adding the flow meter to the system, the flow and pressure of the liquid are disrupted. Usually, the transducer should be small and light compared with the component or process.

Other error sources often encountered are due to human operation, electronic noise, resistance of the wiring and hysteresis.

If the operator of the instrumentation does not understand how to read the visual display or if the display is not clear, then obviously errors can be large. Dial gauges need to be viewed square on to obtain the correct reading. An error due to reading a dial gauge at an angle is called *parallax error*. Think how the speed of a vehicle may appear incorrect when the speedometer is read by the front passenger of a vehicle, as it is read at an angle rather than square on by the driver.

Errors due to electronic noise can be caused by connection wires lying close to other electrical equipment such as a motor or an arc welding set. Magnetic fields from such equipment create small currents in the connection wiring which affect the measurement signal. The measurement signal is usually small so it does not take much interference to create significant errors. The wiring is usually shielded for instrumentation to minimise interference. Electronic filters can be used to allow signals to pass only at the frequency we are interested in, and to block other frequencies.

The resistance of the wiring can sometimes affect measurements. Some transducers use resistance sensing elements. If the resistance of the wiring is significant compared to the resistance of the sensing elements then the accuracy of the instrumentation can be greatly reduced.

Hysteresis occurs in many systems. It refers to instruments that give different readings for the same value of the measured quantity, according to whether the measured value gradually increases or decreases. It usually occurs in mechanical elements due to play or friction in drives. A graph of the instrument output against the measurement input can look like the graph in Figure 7.29. If hysteresis does occur, then its effect should be noted and considered when taking readings.

Errors always exist when setting up any measurement system. Indeed, as instruments get older some components can deteriorate and their characteristics change.

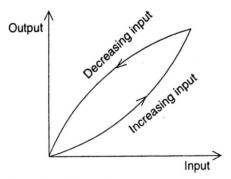

Figure 7.29 Hysteresis error

By carefully considering all of the above errors, these errors can be kept to a minimum and within acceptable limits.

Some other definitions concerning errors

Repeatability is the ability of an instrument to display the same output for repeated inputs of the same measured quantity value. For example, if the resistance thermometer mentioned above was used to take five readings of a constant temperature as follows: 50°C; 51°C; 51.5°C; 50°C; 49°C, then there can be an error with any reading due to a lack of repeatability. The term **precision** is how close repeated readings are to each other. An instrument with a low precision when measuring a constant quantity will produce varying readings scattered over a certain range. An instrument with a high precision will produce similar readings of a constant quantity. Do not get precision confused with accuracy.

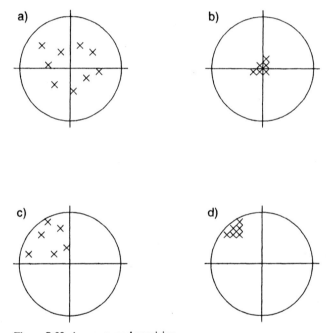

Figure 7.30 Accuracy and precision

A useful comparison to clarify the difference is four dart players throwing several darts at a board (Figure 7.30). The two players of the top two boards, (a) and (b), have a high accuracy while the two players on the bottom, (c) and (d), have low accuracy. The two players on the left though, (a) and (c), have low precision whereas the two players on the right, (b) and (d), have high precision. An instrument with a low precision results in **random errors**. Random errors vary in a random pattern between successive readings. These can sometimes be overcome by taking several readings and calculating an average. Errors that do not vary from one reading to another are called **systematic errors**.

The **reliability** of an instrument is the probability that it will operate correctly to the accuracy specified by the manufacturers.

The **resolution** of an instrument is the smallest change in the measured quantity value that will produce an observable reading at the visual display. If a resistance thermometer has a scale on the display that is subdivided into half degree intervals, then a change in the temperature of 0.1°C cannot be detected accurately. The resolution would then perhaps be 0.5°C.

A **lag** occurs when the quantity being measured takes some time to cause a response at the display output. It is the time that the system takes to respond. Many instruments show a first order response with time when the measured quantity has a step change. This time response needs to be considered or readings will be incorrect.

Sensing elements

The sensing element converts the quantity to be measured into a more suitable corresponding signal. The output from the sensing element can take many forms. A simple example of an instrument sensor is the spring balance. The deflection of the spring is proportional to the force applied. The length of the deflection is measured against a scale. In this way, the measured quantity, the force, is converted into a corresponding signal of length. An electronic measurement system requires the signal to be converted into some sort of electrical signal that can be processed by the signal conditioning unit and then displayed or used for a control system. A mechanically based measurement system transmits the signal as displacement through lever, gears etc. Other types of system use hydraulics or pneumatics. Two different sensing elements are sometimes used. Consider an instrumentation system that measures the flow of water through a pipe. A sensing element can be used to convert the flow rate into a pressure difference. A second sensing element can then be used to convert the pressure difference signal into a corresponding electrical signal. Some examples of sensing elements are as follows.

Resistive sensing elements

Resistive sensing elements convert the measured quantity into electrical resistance and are very common in instrumentation. Copper and nickel are commonly used for sensing elements for **resistance thermometers**. The change in electrical resistance is proportional to the change in temperature.

Strain gauges are used to convert changes in strain into changes in resistance. When metal is stretched, its resistance tends to increase. This is a useful property for constructing strain gauges. These consist of either a wire or foil element that can be stuck onto a surface. Strain is:

$$\text{strain} = \frac{\text{change in length}}{\text{original length}}$$

so when a component is subject to strain then the change in length causes a corresponding change in the electrical resistance of the metal element. A wire strain gauge is made of a length of wire wound in a grid shape. This is supported on a suitable backing material. Strain gauges are also made from foil (Figure 7.31).

Elastic sensing elements are all based on the principle that changes in force or pressure will produce changes in shape or length. The spring balance depends on this principle: the applied force changes the length of the spring. Elastic sensors contain a component that is designed to deform when a force is applied. This deformation can then be measured, with a set of strain gauges for instance.

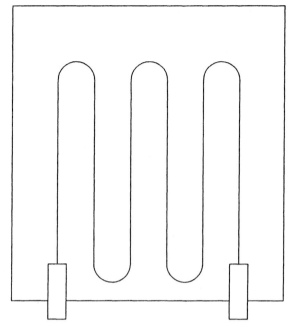

Figure 7.31 Foil strain gauge

A **ring type load cell** is shown in Figure 7.32. The ring deforms as a direct force is applied. The ring can cover a range of loads by using different diameters or thicknesses. Four strain gauges are often used to monitor strain on the inside and outside of each side of the ring. These are incorporated into a circuit. The deflection

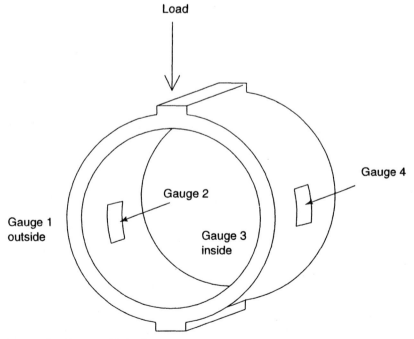

Figure 7.32 Ring type load cell

is then converted into a corresponding change in voltage at the output to the circuit:

change in force → change in deformation → change in voltage

Torque cells operate in a similar manner. They are designed to convert torque into a corresponding deformation of a shaft. Again strain gauges can be used to measure deformation. This is much easier to do if the shaft is stationary. The electrical signal is difficult to transmit from a rotating shaft (Figure 7.33).

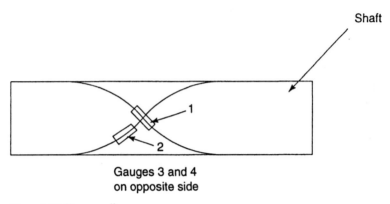

Gauges 3 and 4
on opposite side

Figure 7.33 Torque cell

A Bourdon tube is used as a sensor in pressure gauges. It consists of a tube that is open at one end and closed at the other. The tube is usually in a C-shape although they can be found in other shapes (Figure 7.34). The open end of the tube is

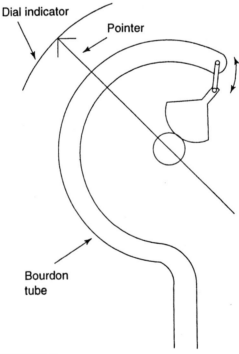

Figure 7.34 Pressure gauge

connected to the process to be measured. As the pressure inside the tube increases, the C-shape opens up by a corresponding amount. Changes in pressure are converted into changes in shape. The end of the C-shape is connected via a gear linkage to a pointer on a dial indicator gauge.

A **photoelectric cell** is used to detect intensity of light. A material is used in the cell so that, when it forms part of an electrical circuit, it can convert changes in the intensity of the light falling on it into a corresponding change in current flowing through it. This takes several forms. One type uses a photoconductive material. The electrical resistance of these materials decreases as the light intensity falling on them increases. This type of sensor is useful when direct contact cannot be made with a process, but changes in light intensity can be related to the quantity to be measured.

Flow measurement

Flow measurement can take three different forms: mass flow rate, volume flow rate or fluid velocity. One common form of the volume flow rate is the **venturi meter**. We have already demonstrated that pressure is proportional to head with the U-tube manometer: $p \propto h$. A venturi meter uses this principle for measuring volumetric flow rate. A narrow throat is arranged in a pipe as shown in Figure 7.35. As the cross-sectional area decreases the velocity increases (see the equation of continuity on page 242). This results in a relationship between volumetric flow rate $\dot{V}$ and the loss of pressure head, $H \, (= h_1 - h_2)$, as follows:

$$\dot{V} \propto \sqrt{H}$$

The reading of H can be arranged against a suitable scale to show volumetric flow rate.

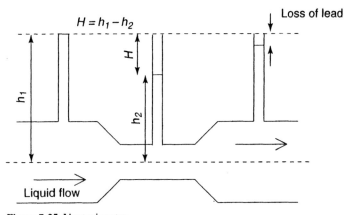

Figure 7.35 Venturi meter

Problems

1. A pressure gauge is specified as having a range of 50–300 kN/m² and an accuracy of ±0.5 kN/m². Explain the significance of this information.
2. A voltmeter has a range of 0–20 V and an accuracy of ±2% full-scale deflection. What is the accuracy for a reading of 7 V?

3. Develop the outline of a control system in block diagram form for maintaining the speed of a diesel engine at a desired value under varying load conditions.

4. A thermocouple is specified with a sensitivity of 0.01 mV/°C. What is the significance of this?

5. A centrifugal governor is to be designed to control the speed of a machine. The shaft of the governor is directly connected to the drive shaft of the machine. There are two weights of the governor each with a mass of 0.1 kg that are to act at a radius of 0.07 m and apply a force to a sensor. Develop a simple mathematical model that describes the relationship between the shaft speed of the machine and the force applied to the sensor by the governor.

6. A water heater holds 15 litres of fluid. The heater is 65% efficient. A model of this plant is required in order to design a simple control system. Develop a simple mathematical model that describes the relationship between the energy input to the system and the temperature changes of the water. Assume that the specific heat capacity of the water is 4.19 kJ/kg K.

7. Define the terms: accuracy, precision, repeatability, sensitivity and reliability.

8. What is the difference between random and systematic errors?

9. What sensors can be used for converting a displacement signal to a voltage signal?

10. Describe the operation of a strain gauge.

11. Draw a diagram showing the major components of an electronic measuring system.

12. What is the resolution of an instrument?

8 Basic electricity

8.1 Basic principles

Electricity is a term that refers to the use of electrical energy in engineering. Electrical energy is so common in modern day life as it is easily transported and easily converted into other forms of useful energy, e.g. for heating, lighting, motors, etc. Electronics is a branch of electrical engineering that involves circuits designed for control, communication and computing. The level of electrical engineering and electronics in modern motor vehicles is becoming increasingly sophisticated and requires specialist knowledge and equipment. You should be aware, though, of the basic concepts involved in the electrical systems of motor vehicles. Some electrical devices are necessary for the operation of the engine and others are designed for the comfort of passengers.

The unit of energy is the joule (J) but electrical energy is commonly measured in *kilowatt hours* (kWh), for historical reasons and scale conveniences.

$$1 \text{ kilowatt hour (kWh)} = 3.6 \times 10^6 \text{ J}$$

The kilowatt hour is the 'unit' of electricity that domestic meters measure. The unit of electrical power is the watt (W). One watt is one joule per second: $1 \text{ W} = 1 \text{ J/s}$. Notice that the unit of energy above, the kilowatt hour, is the unit of power multiplied by time.

Current and voltage

Think of the structure of an atom. At the centre is the nucleus, which is positively charged, consisting of protons and neutrons. Revolving around the nucleus are electrons which are negatively charged. These are the smallest negatively charged particles possible. Some of the electrons of certain materials are only loosely held to the nucleus and they tend to move around at random. These materials are called **electrical conductors**. A material that does not have any loose electrons is called an **insulator**. Most conductors are coated with an insulating material, such as plastics, except at the connection points to prevent accidental connection with other conductors.

Under certain circumstances the electrons of a conductor can be made to flow in one direction, transferring their charge from atom to atom through the material. This flow of electrons is called an electric *current*. The symbol of electric current is I and it is measured in amperes or amps (A). The electric charge of the electrons is measured in **coulombs**. One ampere is one coulomb of charge per second.

For a current to flow in a conductor, a force is required to cause all the electrons to flow in one direction. When a difference in charge exists between two points the electrons will flow from the more negatively charged point to the more positively charged point. The more positively charged point is referred to as the higher potential and the more negatively charged point is called the lower potential. The voltage is a measure of the energy released when a unit of positive charge of one coulomb moves 'downhill' from the higher potential to the lower. The quantity symbol of voltage is V and the unit symbol for the volt is also V. Alternatively voltage is a measure of the work done in moving a charge 'uphill' from the lower potential to the higher potential. Voltage is sometimes called potential difference or electromotive force (EMF). When a current of one ampere flows between two points and one watt of power is dissipated between the two points, then a potential difference of one volt exists between the two points.

Notice that the wording above is 'from the more negatively charged point to the more positively charged point' and not from the negative point to the positive point. This is because electrons will flow even if the two points were, say, positive but one was more positive than the other. Notice also that the voltage is measured *between* or *across* two points and that current is measured *through* a conductor. Never refer to a voltage through a circuit as this is complete nonsense. A current flows through something because a voltage is placed across it. This is further demonstrated by the way that voltmeters and ammeters are applied to circuits. A voltmeter measures voltage. An ammeter measures current.

Look at Figures 8.1 and 8.2. The voltmeter is connected across the device to measure the potential difference between the two points. The ammeter is connected as a conductor in the circuit to measure the current flowing through it.

When electrons flow through a conductor in one direction, the current is referred to as direct current (DC). If the direction is reversed at frequent intervals then this is referred to as alternating current (AC). Domestic and industrial electricity supplies are generally alternating current as this is easier to generate and transmit. In motor vehicles, direct current is used as this is more suitable for use with a battery. The theories involving direct current are simpler than those involving alternating current, as quantities are not time-dependent.

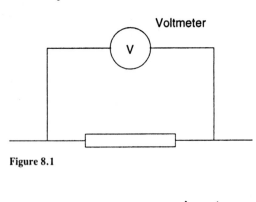

Figure 8.1

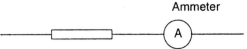

Figure 8.2

Circuits and resistance

When a voltage is applied to a conductor, the current that flows depends on the magnitude of the voltage, the dimensions of the conductor and the material of the conductor. The dimensions and the material type are summarised by the resistance. Resistance, as you may expect, is the resistance to current flow. The quantity symbol of resistance is R and it is measured in ohms (Ω). The greater the resistance of a conductor the less current will flow for a given voltage. Generally, the resistance of a conductor of a given material increases with decreased cross-sectional area and increased length. The relationship between voltage V current I and resistance R is:

$$R = \frac{V}{I}$$

This is known as Ohm's law.

❏ Example 8.1

A voltage of 10 V is applied across a lamp of resistance 5 Ω. Calculate the current that flows.

If $R = \dfrac{V}{I}$,

then $I = \dfrac{V}{R}$

$$= \frac{10\,[\text{V}]}{5\,[\Omega]} = \mathbf{2\,A}$$

Before an electrical current can flow, the voltage must be applied to a complete circuit of conductors. If the voltage is applied to conductors that do not form a closed circuit, the circuit is referred to as an **open circuit**. A device that is part of a circuit such as a light or a motor is referred to as a **load**. The load forms a resistance in the circuit. Usually, when circuits are being analysed, the conductors, i.e. the wiring, are assumed to have zero resistance and all the circuit resistance is due to the load. This is not actually the case but the resistance of the conductors is negligible compared with that of the load and calculations are a lot easier. If the load is bypassed by a conductor, then this is referred to as a **short circuit**. If the resistance of the conductor is considered to be zero then, if we apply Ohm's law, it implies that an infinitely large current will flow. This is not the case in practice as nothing has zero resistance but, as anyone who has dropped a spanner across the terminals of a car battery will know, a very large current can flow resulting in serious damage. For this reason, fuses or circuit breakers are fitted in circuits. A fuse is a 'weak' point in a circuit, designed to burn out when a certain maximum current flows and before damage can be done to the rest of the circuit. A circuit breaker is designed to open, breaking the circuit before a current gets too large to cause damage.

There are three important rules concerning voltage and current:

(a) At any junction in a circuit, the sum of the current in is equal to the sum of the currents out.

(b) The voltage drop around a closed circuit is zero.
(c) The power P that a load takes in a DC circuit is calculated from the current multiplied by voltage: $P = VI$.

These will become clear in a moment as we look at an example.

The direction of current flow can cause some confusion. As has already been stated, current flow is the flow of electrons from the more negative point to the more positive point. Also above, the movement of a positive charge was mentioned in the discussion on current. When electricity was first discovered, the flow of electrons was not understood and it was assumed that current flowed from more positive to more negative points in a circuit. Many rules and theories were developed on the assumption that current flow was in this direction. Later, it was discovered that electron flow was from the more negative to the more positive points. Rather than change all the rules and theories, the convention of current flow being from positive to negative has been maintained. Remember what current flow actually is, but in any calculations always assume that current flow is from the more positive point to the more negative one.

When a circuit consists of several resistors, it is usual to find the overall effect they have on the circuit and consider them as one resistance. Resistors that are in series around a circuit can be directly added together.

☐ *Example 8.2*

Calculate the current that flows around the circuit shown in Figure 8.3. Find the voltage across each resistance and the overall power consumed.

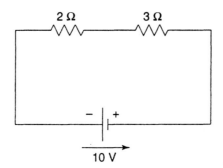

2 Ω   3 Ω

10 V

Figure 8.3

Notice the symbols used for the resistors and the cell battery. These are standard symbols used for circuit diagrams.

Total resistance, $R = 2\,[\Omega] + 3\,[\Omega] = 5\,[\Omega]$

$$I = \frac{V}{R}$$

$$= \frac{10\,[V]}{5\,[\Omega]} = 2\,A$$

The voltage across each resistor can be found by applying Ohm's law again.

Voltage across 2 Ω resistor

If $R = \dfrac{V}{I}$,

then $V = RI$
$= 2\,[\Omega] \times 2\,[\text{A}] = \textbf{4 V}$

Voltage across 3 Ω resistor
$V = RI$
$= 3\,[\Omega] \times 2\,[\text{A}] = \textbf{6 V}$

The voltages across loads are referred to as volt drops, since they act in the opposite direction to the current flow. If the supply voltage is considered as positive around the circuit then the volt drops will be negative. This demonstrates law (b) above, that the total voltage around the circuit is zero:

$$10\,[\text{V}] + (-6\,[\text{V}]) + -(4\,[\text{V}]) = 0$$

The power consumed can be found from law (c).

$P = VI$
$= 10\,[\text{V}] \times 2\,[\text{A}] = \textbf{20 W}$

Note that this law applies only for DC circuits.

When a circuit does not work, a voltmeter can be applied to any load. If the voltmeter shows a zero voltage drop across the load, then this indicates that there cannot be any current flowing and there is a break in the circuit, i.e. a loose connection. By applying the voltmeter across various points along the circuit, the total terminal voltage can be measured across the break in the circuit.

Example 8.3

Calculate the current that flows through the main conductors in the circuit in Figure 8.4. Calculate the current in each branch of the circuit.

When resistors are arranged like this they are said to be in parallel. Resistors in parallel cannot be added directly together.

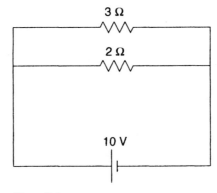

Figure 8.4

The relationship between the overall resistance and n individual resistances is as follows:

$$\frac{1}{R} = \frac{1}{R_1} + \frac{1}{R_2} + \ldots + \frac{1}{R_n}$$

$$\frac{1}{R} = \frac{1}{2} + \frac{1}{3}$$

$$\frac{1}{R} = \frac{3+2}{6}$$

$$\therefore R = \frac{6}{5} = 1.2\,\Omega$$

The voltage applied across each of the two resistors in parallel is the supply voltage. The current in the 3 Ω resistor branch:

$$I = \frac{V}{R} = \frac{10\,[V]}{3\,[\Omega]} = 3.333\,A$$

The current in the 2 $[\Omega]$ resistor branch:

$$I = \frac{V}{R} = \frac{10\,[V]}{2\,[\Omega]} = 5\,A$$

Total current in main conductor:

$$I = \frac{V}{R} = \frac{10\,[V]}{1.2\,[\Omega]} = 8.333\,A$$

Notice that this verifies law (a):

$$5\,[A] + 3.333\,[A] = 8.333\,A$$

Another term you may come across is **conductance**. In DC circuits conductance is the reciprocal of resistance. The quantity symbol is G and the unit is the siemens (S).

$$G\,[S] = \frac{1}{R\,[\Omega]}$$

8.2 The effects of electric current

Electric currents are created in electrical circuits usually for one of three reasons:

(a) a heating effect
(b) a magnetic effect
(c) a chemical effect.

Heating effect

When current flows through a material there is a heating effect. This effect can be small in a well designed circuit. Loose connections tend to cause some heating. Some circuits are designed to create a large heating effect such as in an electric heater or in a light bulb where the filament is heated to a temperature where it glows to give out light energy. If the heating effect is too great for the design of the circuit then the wiring can become too hot and the insulation will break down.

Magnetism

Magnetism can be created in two ways: either by permanent magnets or by an electric current. If a conducting wire is wound into a coil then the magnetic field produced will be similar to that of a permanent magnet. The strength of the field is proportional to the number of turns of wire and to the current flowing through it. If the coil of wire is wound onto an iron core, then the magnetic effect is increased since iron provides a path for the magnetic field lines.

The arrangement in Figure 8.5 is known as an electromagnet. Its advantages over permanent magnets are that large magnetic forces can easily be created and that it can be switched on and off. If a conductor travels through a magnetic field then a potential difference is produced in that conductor. This is known as Faraday's effect. The voltage produced is due to the relative movement between the magnetic field and the coil. If the conductor forms part of a circuit then the voltage will cause a current to flow.

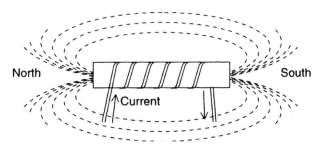

Figure 8.5

This effect is very useful and is used to generate electrical energy from mechanical energy (see Figure 8.6). The dynamo or alternator of a car generates electricity in this manner, as do all commercial generators in power stations. If the reverse is now applied, and a current made to flow through a conductor in a magnetic field, then this produces relative movement between the conductor and the magnetic field. This is the principle of the electric motor. An electric motor is the same as a generator except that electrical energy is supplied for conversion to mechanical energy rather than the reverse. In an electric motor an electric current is made to flow through a conductor in a magnetic field. The magnetic field can be caused by permanent magnets or electromagnets.

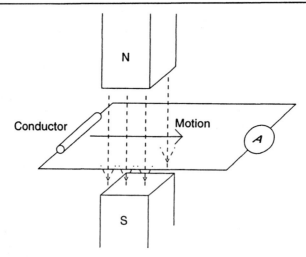

Figure 8.6

Chemical effects

When a current flows through certain liquids, chemical changes take place to the liquid and to the metal placed in the liquid. Electrical energy is converted into chemical energy. One of the disadvantages of electrical energy is that it is difficult to store directly. In a motor vehicle electrical energy is generated continuously by the fan belt driving a generator. However, the electrical power required for turning over an engine and starting a vehicle is vast, and this electrical power needs to be stored ready for when the vehicle is next started. In a battery, energy can be stored as chemical energy for later use.

A car battery consists of a container containing several 2 V cells (six for a 12 V battery). In each cell, two different types of lead plate are immersed in an electrolyte solution of sulphuric acid and distilled water. A voltage applied to the battery of cells creates an imbalance of electrons and a current passes through the electrolyte. The positive plate has a deficiency of electrons and the negative terminal has an excess of electrons. As the current flows through the electrolyte, chemical reactions take place between the plates and the electrolyte and a voltage is maintained across the plates. This is the charging of the battery.

When electrical energy is later required, the chemical changes are reversed and when the battery is part of a circuit a current can flow again. This is the discharging of the battery. The battery can be charged and discharged many times during its life, but if it is completely discharged (flattened), for example by leaving the headlights on overnight, then its life can be shortened. A battery can sometimes be accidentally completely discharged if the electrical circuitry of a vehicle becomes wet. Impure water conducts electricity so water that connects a more positive point of the circuit with a more negative point will allow a current to flow and can gradually discharge the battery.

A typical charging circuit is shown in Figure 8.7. The generator is driven by the fan belt to produce electrical energy. The regulator limits the maximum voltage that the generator can produce to avoid damage to itself and to other equipment. The cut-out

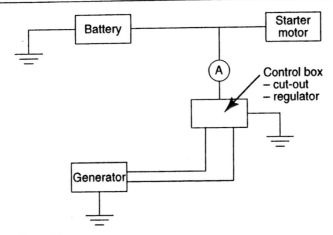

Figure 8.7

is a one way switch that allows the battery to be charged by the generator but prevents the battery discharging through the generator.

The symbol consisting of three reducing parallel lines represents what is termed the **earth** of the vehicle. The frame of the vehicle is used as one big terminal for the electrical system and this is the earth of the vehicle. One terminal of the battery is designated as the earth and is directly connected to the vehicle frame. This is done to simplify wiring and to save on cable. For example, notice that both the battery and the generator are connected directly to the earth, i.e. to the vehicle frame. If the frame were not used as the earth another cable would be required between the generator and the battery.

Problems 8

1. When a circuit is tested a current of 0.5 A passes when a voltage of 24 V is applied. The total load resistance of the circuit is 40 Ω. What is the resistance of the rest of the circuit?
2. Three loads of resistance 3 Ω, 6 Ω and 9 Ω are connected in series and a voltage of 12 V applied. What current passes through each resistor?
3. Three loads of resistance 2 Ω, 5 Ω and 8 Ω are connected in parallel and a voltage of 24 V applied. What is the current that passes through each resistor?
4. Three loads of resistance 4 Ω, 1 Ω and 7 Ω are connected in parallel. A further resistance of 10 Ω is connected in series with them. What is the total resistance of the circuit?
5. Three 12 V headlights are to be connected to a 12 V battery. How would you connect the circuit up to achieve the correct voltage across each?
6. An electric windscreen heater requires 30 W of power. When the windscreen heater is connected to a 12 V supply, what current passes through the circuit?
7. Show how you would connect up a voltmeter and an ammeter to test a circuit.
8. What is the purpose of a vehicle battery when a generator is already used?
9. The current passing through a simple circuit is found to be too low. Why and how might the conduction leads be changed to improve this?
10. Describe briefly the three effects of an electrical current in electrical circuits.

9 Metals

9.1 Steel

A wide range of materials is used in modern engineering. Materials can be broadly classified into two groups: metals and non-metals. Non-metals include plastics. Plastics are being used increasingly in engineering. As technology advances, more plastics are able to be tailor-made to have specific properties to suit a particular job. In a motor vehicle, however, the main construction material is still metal. This chapter will concentrate on the properties of metals and their uses. Common types of metals are **ferrous metals** which contain iron. These tend to be cheaper than **non-ferrous metals** and their structure and properties can easily be changed by heat treatment and the addition of alloying materials. Non-ferrous metals are sometimes used, because of their specific properties, for example: being good conductors of heat and electricity, or having a low coefficient of friction with other surfaces.

Mechanical properties

Strength is the ability to stand applied forces without breaking, or the resistance of a metal to stress. Tests are carried out on a material sample of standard dimensions (see page 71). Strength can be defined in several ways:

- **tensile strength** is the maximum force per unit area that the material will withstand when loaded in tension
- **compressive strength** is the maximum force per unit area that the material will withstand when loaded in compression
- **shear strength** is the maximum force per unit area that the material will withstand when loaded with a shear force
- **yield strength** is the force per unit area at which the material gives way or yields.

It is important that a material is not loaded to the yield point. After this has occurred the material takes on a permanent set and the characteristics change, making it unsuitable for further load carrying.

    Toughness is the ability of a metal to resist impact loads without fracture. A chisel shank that has to withstand hammer blows needs to be tough. Toughness can be measured by an impact test such as the **Izod test**. Here, a test piece of material gripped in a vice is hit with a hammer attached to a pendulum. The force of the blows applied can then be made equal by swinging the pendulum from the same height. The hammer and pendulum arrangement is designed to break the test piece. The effect of the blow on the test piece is measured by the height that the pendulum reaches as it

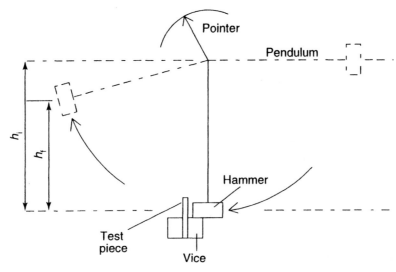

Figure 9.1 Izod test

continues to swing after impact. The height that the pendulum swings short of its original height is a measure of the energy lost at impact (Figure 9.1).

If the initial height of the pendulum is h_i and the final height is h_f then the loss of energy due to the impact can be represented by the loss of potential energy:

$$mgh_i - mgh_f = mg(h_i - h_f)$$

The energy lost at impact is recorded by a pointer that moves to the height reached by the pendulum against a scale. Since the energy lost is directly proportional to the lost height, the scale can be calibrated to measure joules.

Hardness is the ability of a metal to withstand wear and scratching. Any bearing surfaces need to be hard to withstand wear. Tools such as files need to have hard surfaces so the teeth will cut the work material and not wear out quickly.

Ductility is the ability of a cold metal to be stretched and formed without breaking. The metal used for sheet metal work needs to be ductile, so the components can be shaped from a flat sheet. A ductile metal, copper for example, can be *drawn* through a die to make wire without breaking. Ductility can be measured as a percentage of elongation at fracture. Two marks are made on a test piece of material a certain distance apart. The test piece is then stretched until it breaks. The two pieces are then placed back together and the distance between the two marks measured again. The difference between the original length between the two marks and the new distance apart is the extension:

$$\text{percentage elongation} = \frac{\text{extension}}{\text{original length}} \times 100\%$$

Malleability is sometimes confused with ductility. Malleability is a term that describes the ability of a metal to be shaped by **compression** loads. It is therefore a measure of how easily the metal can be shaped by hammer blows for example. Some materials, lead for example, can be malleable without being ductile.

Elasticity is the ability of a material to return to its original shape after it has been deformed by a load.

Besides these mechanical properties, there may be other physical properties of interest to an engineer when selecting materials, such as thermal conductivity and electrical conductivity.

Ferrous metals

The term ferrous means iron. A ferrous metal therefore means a metal that includes iron. Pure iron is not usually used in engineering as it is soft and not very strong. Iron is used to make alloys. Steel and cast iron are alloys of iron and carbon. An alloy is a mixture of two or more elements, the main one being a metallic element. Steel contains less than 1.8% carbon. Cast iron contains between 2 and 4% carbon. When the carbon content is below 1.8% all the carbon forms chemical compounds with the iron. These compounds enable the characteristics of the steel to be altered by heat treatment. When the carbon content is above 1.8%, as with cast iron, pure carbon exists in the iron.

Steel is manufactured from iron ore. The first process is the conversion of iron ore to pig iron. To do this iron ore, coke, and limestone are heated in a blast furnace. Hot air is forced up through the materials, making the coke burn rapidly and reducing the iron ore to molten iron. The molten iron absorbs carbon as the coke is burnt. The limestone combines with impurities in the iron ore to form slag. The slag can be separated from the iron. The result is pig iron. Pig iron contains too much carbon and other impurities. The molten pig iron is mixed with scrap iron, and then undergoes one of several other processes that remove the impurities by combining them with oxygen. After the impurities have been removed, small amounts of other materials may be added to alter the properties of the resulting steel. The carbon content can also be adjusted at this point.

The amount of carbon in plain carbon steel determines its properties. These properties can be changed by heat treatment. The three most important properties to an engineer are hardness, ductility and strength. In its normal state (i.e. without any form of heat treatment), as the carbon content of steel increases the hardness increases and the ductility decreases. The strength of the steel increases as the carbon content increases up to a maximum when the content is 0.83%. Above this, the strength reduces. Low carbon steel is known as mild steel and has a carbon content of about 0.2%. Medium carbon steel has a carbon content of between 0.4 and 0.6%. High carbon steel usually refers to steel where a maximum strength is required when the carbon content is 0.83%. Cast iron contains more than 2% carbon. This causes a low tensile strength, and cast iron components should only carry compressive loads. It is brittle and cannot stand impact loads either. The high carbon content of cast iron does however make machining easy, and its low melting point means it is easy to cast.

Uses of carbon steel

Mild steel is a very common material in engineering. It is usually manufactured in bars, sheet form or plate form. Machining is carried out on the steel in the bar form. Sheet steel is used for pressing shapes. The plate form is used for pressure vessels such as boilers. Mild steel does not alter significantly with heat treatment due to its low carbon content.

Medium carbon steel has a higher tensile strength than mild steel. It is commonly used for hand tools. The carbon content allows heat treatment to be applied in order to increase hardness. It cannot be completely hardened but it can be toughened effectively, and is commonly used for keys. High carbon steel can be made very hard with heat treatment, and is used for cutting tools such as files. Cast iron is used to economise on material when making a component. A shape can be easily cast, such as an engine block, and then only the necessary surfaces machined to the required shape and finish. The low tensile strength needs to be considered when designing the part, and dimensions need to be large enough to make up for this.

Impurities in steels

Not all impurities can be removed from steel during manufacture. The impurities must be controlled to produce steel with the desired properties. The main impurities are sulphur, manganese, phosphorus and silicon. Sulphur causes the steel to be brittle at high temperatures. The steel is then difficult to forge. Manganese tends to combine with unwanted impurities, such as sulphur, in molten steel, which are then removed as slag. Any remaining manganese hardens the steel. Phosphorus reduces the toughness of carbon steels. The content must be controlled to within acceptable limits. Silicon can increase the strength of steel, but it does not do this as well as carbon and it can reduce the effectiveness of carbon.

Alloy steels

An alloy steel is a steel with another element or alloy added to alter its properties in some way.

Manganese steel contains more than 1% manganese. As mentioned previously, manganese increases the hardness of the steel and so is useful for resisting abrasion.

Tungsten is added to steel, forming tungsten steel, to increase the strength, hardness and toughness. These properties are maintained at high temperatures. This steel is commonly used in high speed steels (cutting tools).

Molybdenum has a similar effect on steel to tungsten, maintaining the hardness at high temperatures.

When **nickel** is added to steel the tensile strength and the toughness are increased. The hardness and corrosion resistance are also increased.

Chromium is added to steel to increase the hardness.

Nickel and chromium are sometimes used in combination to form **nickel-chromium steel**. The resulting mechanical properties are: increased elasticity, greater hardness and greater toughness; the resistance to corrosion is also greatly increased. Many stainless steels are based on nickel-chromium steel. A common form of stainless steel is made with 18% chromium, 8% nickel and 0.08% carbon and it has a high resistance to corrosion and staining.

Cobalt is used to make various alloy steels, typically containing 5–10% cobalt, with tungsten and chromium. This forms hard but brittle steels in high speed tools.

9.2 Heat treatment

If steel containing sufficient carbon is heated to a sufficiently high temperature and then allowed to cool, changes occur in the structure which alter the mechanical properties. These changes to the structure can be controlled by cooling the steel at

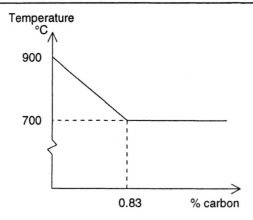

Figure 9.2 Heat treatment temperatures

different rates through different temperature ranges. If the steel is cooled rapidly, the changes that occur during the heating process do not have time to reverse and the steel becomes very hard. The steel can be cooled more slowly by quenching it in oil, reversing some of the changes and making the steel very tough. If the steel is cooled very slowly, all changes can be reversed and the steel is soft. The temperature to which the steel must be heated to make these changes take place depends on the carbon content. These are shown in Figure 9.2.

The necessary heating temperature for a particular carbon content is called the **critical temperature**. The steel is usually heated to around 25°C above the critical temperature to ensure all the necessary changes have taken place in the structure of the steel. The temperature is a maximum for pure iron at 900°C reducing to 700°C for 0.83% carbon steel. With more than 0.83% carbon the critical temperature remains constant at 700°C.

Hardening

Steel can be hardened by heating it to the critical temperature. The steel is then cooled rapidly by plunging it into cold water. This is not very effective when the carbon content is below 0.25%.

Tempering

After a piece of steel has been hardened it is usually too brittle to be useful as a tool or component. Some of this hard material can be broken down and the toughness increased by reheating the steel. This process is called *tempering*. The amount of hardness removed depends on the temperature that the steel is heated to: the higher the temperature the more hardness is removed. For example, a cutting tool that must be very hard with just a little of the brittleness removed, is tempered by heating to around 230°C. A tool that has to withstand hammer blows is tempered to a higher temperature to make it tougher. A chisel would be tempered at around 280°C. As the temperature of the steel changes during tempering, oxides on the surface of the steel will change colour. This can be used as an approximate guide to the temperature for simple tempering jobs.

Annealing

Annealing is the process of making the steel as soft as possible. The steel is heated to above the critical temperature and then cooled as slowly as possible. This is best done by heating in a furnace and then turning the furnace off to cool with the steel still inside.

Normalising

Normalising is regarded as a corrective treatment when a steel has become hardened after undergoing a manufacturing process of some sort. The steel is softened by heating it to above the critical temperature and then allowed to cool in air at room temperature.

Case hardening

Some engineering components need to be tough to absorb impact loads but also require a hard surface to resist wear. This is achieved by making the component out of a steel that has the necessary toughness and then case hardening it. Case hardening is a process that increases the carbon content of the outer layers of the component, by heating it with other carbon-rich materials. The outer layers of the component become hard high carbon steel while the inner section remains in its original form. Gudgeon pins are fabricated in this manner. The piston imposes high loads on the gudgeon pin requiring a tough material, but the angular motion of the connecting rod requires a wear-resistant surface.

9.3 Non-ferrous metals

Copper is a very good conductor of heat and electricity. Thin wire and sheet are easily manufactured because copper is very ductile. For this reason, copper is used extensively for electrical wiring and contacts, and for heating system materials. It is too soft to be used in its pure state for many applications, and is usually used in an alloy form of either brass or bronze. Brass is an alloy of copper and zinc. The properties can be altered by varying the proportion of zinc between 30% and 40%. Bronze is an alloy of copper and tin. A common form is phosphor bronze, which is about 88% copper, 10% tin, and has 0.25% phosphor added. Phosphor-bronze is often used for bearing bushes. Another common form is called gunmetal which is about 88% copper, 10% tin, and has 2% zinc added. Gunmetal has a high resistance to corrosion and makes very strong castings for such things as pump housings. A tube and fin radiator of a cooling system typically has brass tubes and a brass header tank, with copper fins.

Aluminium is a good conductor of electricity; it is very light, and has good corrosion resistance. As with copper, it is too soft to be used in its pure form where strength is required. It is used for high voltage overhead electricity cables because of its lightness and good conductivity. These cables have a steel core because aluminium is not strong enough to support its own weight. Aluminium is used to make lightweight alloys, usually, for sheeting, or for castings. Composite pistons are constructed from a combination of steel and aluminium, to take advantage of the light weight and high thermal conductivity of aluminium and the strength of steel.

White metal is an alloy used as a bearing material. The metal antimony is combined with lead or tin. The antimony is the harder material that resists wear, and the lead or the tin encourages a film of lubricant to be retained when the metal is used as a bearing surface. Most engines used in motor vehicles use tin-based white metal inserts which fit inside a shell. This makes replacing the bearing simple and cheap.

9.4 Corrosion

Corrosion of metals is caused by chemical attack. Corrosion of metal components causes damage to the surfaces. It reduces the mass of the original metal and affects the mechanical properties such as strength.

The two major causes of corrosion are:

1. **rusting** which is the reaction of iron with oxygen in the presence of water to form an iron oxide
2. dissimilar metals in contact in the presence of water.

Steel is particularly vulnerable to corrosion where there are impurities in the steel and salts in the water, which is why vehicle body corrosion is worse at the coast or in winter when salt is spread on roads. To avoid oxygen and water, steels are usually coated with another material such as paint or a corrosion-resistant metal.

The surfaces of zinc, aluminium and magnesium oxidise very rapidly. The oxide surfaces formed are extremely corrosion-resistant and no further oxidation will take place. Unfortunately this is not the case with steels, since the rust formed gives no protection. When two dissimilar metals are in contact in the presence of moisture, a current can be set up in the similar manner to the operation of a cell battery. A chemical reaction called **electrolysis** takes place; the liquid is called an **electrolyte**. As the current flows, electrons are transferred at the metal surfaces; these are called **electrodes** when referring to electrolysis. One of the electrodes will gain electrons from the electrolyte. The atoms of the other electrode can lose electrons and form **ions** which go into solution, resulting in electrochemical attack on the surface of the metal. Care must be taken when choosing metals for construction to avoid electrolytic action. Particularly serious is the combination of copper and aluminium. Aluminium alloy sheeting joined together with copper rivets would be disastrous because the aluminium would corrode at an extremely high rate.

Problems 9

1. Describe the difference in composition between cast iron and steel.
2. Under what conditions should cast iron not be used? For what purposes might cast iron be used?
3. Describe four heat treatment processes that are applied to steel to alter its mechanical properties?

4. Steel with a carbon content of 0.8% is fully hardened and is to be used for cutting tools. It is found to be too brittle for this purpose. Describe the heat treatment process, including temperatures, that could be applied to reduce the brittleness.
5. What is an alloy steel? What is stainless steel and where might it be used?
6. Aluminium is soft and weak. Why then is it used in engineering? How might the weakness of the material be overcome?
7. What property of copper makes it easily drawn into wire?
8. What is brass and what is bronze?
9. Why is corrosion of steel faster in a salt water environment?
10. Describe how a tough steel component can be given a hardwearing surface without affecting the properties of the inner section of the material.

Problem answers

Chapter 1   Introduction

1) $90.922 \times 10^{-3}\,m^3$, 90.922 litres
2) $1.2 \times 10^{-3}\,m^3$
3) 113.398 kg
4) 1.6 m
5) 809.205 kg
6) 9.982 m
7) $3.978 \times 10^{-3}\,m^3$, 3.978 litres
8) 2.319 m
9) Yes, 0.006 inch = 0.1524 mm
10) 17.882 m/s
11) 1.065
12) 55.920
13) 1.057×10^{21}
14) $S = \dfrac{v^2 - u^2}{2a}$
15) $r = \dfrac{v^2.2.h}{S.g}$
16) 2.45 m
17) 58.6°, 31.4°
18) 0.34 m, 0.73 m, 65°
19) 35°
20) y = 454

Chapter 2   Forces

1) 3.924 kN
2) 39.375 kg
3) $2.871 \times 10^{-3}\ m^3$
4) 365.9 kg
5) 350 bottles
6) $0.295\ m/s^2$
7) 841.7 N
8) $2.702 \times 10^{-3}\,m^3$ or 2.7 litres
9) $866.67\ m/s^2$
10) $0.07\ m^3$
11) 3.679 kN
12) 24.057 kN, 12.258 kN
13) 984.8
14) 13.042 kN, 2.300 kN
15) 2.266 kN, 1.133 kN
16) 8.886 kN
17) —
18) 16.85 kN, 11.21 kN
19) 1.763 kN, 10.154 kN
20) 5.096 kN, 0.983 kN, 11°
21) 52.50 Nm
22) 175 N
23) 333.9 mm from the left end
24) 323.077 N
25) 2.117 m from the front axle
26) 4700 N
27) 78 mm to the rear
28) 542 mm from the left end
29) 114.95 mm from the left end
30) 2.4 mm to the right

Chapter 3   Distortion of materials

1) $31.124\ MN/m^2$
2) $191\ GN/m^2$
3) $127.324\ MN/m^2$
4) $22.222\ MN/m^2$
5) —
6) 3.183 J
7) $28.937\ MN/m^2$
8) 25 kN, 8.75 kNm
9) 29.167 kN, 12.5 kN, 16.667 kN
10) 24.375 kNm
11) $F_{copper} = 169.7\ N$, $F_{steel} = 124.6\ N$
12) 13.7 kN
13) 0.625 mm
14) 17
15) $450\ MN/m^2$
16) 4 m
17) 235.9 kNm
18) 40.709 mm
19) 360 N, 240 N
20) 6.425 kNm
21) $64.236\ MN/m^2$

Chapter 4   Motion

1) 3.6 s
2) 6.9 s

3) 2.4 s, 6.9 m/s^2
4) 2222.24 m
5) 73.7 km/h at 241.3° or 28.7° west of south
6) 477.5 revolutions
7) 111.1 rad/s
8) 0.3 m/s due north
9) 6.667 km/h
10) 0.2, 1.2 N
11) 441.254 N
12) 13.5
13) 20.8°
14) —
15) 6621.75 N
16) 82.375 N
17) 86.280 N, 54.535 N, 63.230 N
18) 0.26
19) 2.158 Nm
20) 0.174
21) 260.61 Nm, 40.936 kW
22) 10 966.227 m/s^2
23) 98.9 km/h
24) 21.5°, 61.8 km/h
25) 2.51 m at 31° anticlockwise from A
26) 55 m
27) 444.1 m/s^2

Chapter 5   Work, energy and power

1) 360 W
2) 24 525 J
3) 277.78 kW
4) 20.9 kW
5) 8.886 MJ
6) 117.3 kJ
7) 1123.8 J
8) 12.984 kW
9) 89.4%
10) 132.5 mm
11) 8.16
12) 2.7 kN

Chapter 6   Thermodynamics

1) 294 K
2) 281.04 kJ
3) 2.197 bar
4) 108.9 kN/m^2

5) 12 972 kJ
6) 6140.3 kJ
7) 2050 kJ from the system
8) 90 kJ
9) 166.67 kN/m^2
10) 1545°C
11) 32.9 W/m^2
12) rough surfaces
13) 51.7 g
14) 18.120 m/min, 7.707 m/min
15) 7.007 W
16) 0.63 m/s
17) 16 567.028 kN/m^2
18) 58.5°C
19) 952 kg/m^3
20) 5.8 m/s

Chapter 7   Control and instrumentation

1) —
2) 6.6 – 7.4 v
3) —
4) —
5) $F = 0.014 \, \omega^2$, where F is in N and ω is in rads/s
6) $Q = 96.7 \, \Delta T$, where Q is in kJ and ΔT is in K
7) —
8) —
9) —
10) —
11) —
12) —

Chapter 8   Electricity

1) 8 Ω
2) 0.667 A through each
3) 12 A, 4.8 A, 3 A
4) 10.718 Ω
5) In parallel
6) 2.5 A
7) Voltmeter – across circuit (in parallel), ammeter – in line in circuit (series)
8) To store electrical energy for starting the engine

9) Change to leads with a larger
C.S.A.
10) —

Chapter 9   Metals

1) —
2) Not for tensile or impact loads

3) —
4) —
5) —
6) —
7) Ductility
8) —
9) —
10) —

Index

Printed in the United Kingdom
by Lightning Source UK Ltd.
105905UKS00002B/45

9 780340 645277